A Bibliography of Noise

for 1974

A Bibliography of Noise
for 1974

Compiled by

Judith Kramer-Greene

The Whitston Publishing Company

Troy, New York

1976

Copyright 1976 by
Judith Kramer-Greene
Library of Congress Catalog Card Number: 72-87107
ISBN 0-87875-078-9
Printed in the United States of America

PREFACE

This is the fourth annual volume documenting nearly all the literature surrounding noise for the previous year. It is an attempt at a near-complete world bibliography, for 1974, of books and periodical literature and is designed as a fourth supplement to the basic volume, A BIBLIOGRAPHY OF NOISE, 1966-1970 (Whitston, 1972). The bibliography is divided into two sections: a title section arranged alphabetically; and a subject section, arranged alphabetically by subject and alphabetically by title within subjects. Titles and subjects are provided so that researchers who do not trust subject assignments, which are always arbitrary and unique to the compiler, can conduct their own to-some-extent original literature search. Especial care has been taken to develop subject headings meaningful to the user: entries are broken down into almost 212 subject categories, and they have been allowed to arise from the nature of the material, and not, that is, imposed on the matter by external lists such as that provided by the Library of Congress, et al. In addition, adequate "See" and "See Also" entries are included.

The bibliography is designed to serve the needs of different kinds of researchers, from medical scientists to those with only passing interest in this increasingly severe social problem: it treats noise and its physiological, sociological, and cultural effects covering the previous year, and it should be of service to the undergraduate, graduate, and professional student of medicine, psychology, sociology, and the physical and applied sciences; the architect, the educator, and the musician, among others.

The following bibliographies, serials indexes, and abstracts have been searched in compiling this bibliography: *Applied Science and Technology Index; Art Index; Bibliographic Index; Biological Abstracts; Books in Print; British Books in Print; British Humanities Index; Business Periodicals Index; Canadian Periodical Index; Catholic Periodicals & Literature Index; Cumulative Book Index; Current Index to Journals in Education; Education Index; Hospital Literature Index; Index Medicus; Index to Legal Periodicals; Index to Nursing Literature; Index to Periodical Articles Related Law; Index to Religious Periodical Literature; International Nursing Index; Law Review*

Digest; Library of Congress Catalog: Books: Subjects; The Music Index; Philosophers Index; Public Affairs Information Service; Readers Guide to Periodical Literature; Social Sciences and Humanities Index; and Whitaker's Cumulative Book List.

In using the subject section of this bibliography, one should note the following reminders: 1. Noise as vibration is generally not included unless it involves the physiological, and psychological, or the emotional. 2. Noise-induced pulsations, for example in oil burners, are not treated. 3. No attention has been paid to acoustic stresses to metals, generally. 4. No engine noise, as in the instance of an internal combustion engine, has been included when that noise pertains purely to engine or metal fatigue. 5. White noise, except in six instances, has been excluded. 6. Although industrial noise and machinery noise have, generically, their subject heads, they are also broken down into specific sub-components. 7. "Children," "Neonatal," and "Youth" are separate subject heads. 8. Countries are listed: "Noise: Canada," "Noise: Germany," and so forth, with states listed alphabetically under "Noise: United States."

Thus, this bibliography concerns noise and its effects on people, their culture, and cultural artifacts, such as architecture, cities, music, vehicles, whether automobiles or airplanes, and the like. It is concerned with acoustic stress on man, the animal, his physical and mental and emotional well-being. And thus it must be concerned with the noise of transformers and air-conditioners, with that of apartments, of dental drills, and of rock bands; but not with, for example, most forms of radio noise, white noise, and masking noise.

LIST OF PERIODICAL ABBREVIATIONS

ABBREVIATIONS	TITLE
ASHRAE J	ASHRAE Journal (New York)
ASSE J	American Society of Safety Engineers (Park Ridge, Illinois)
ASTM Stand N	ASTM Standardization News (Philadelphia)
Acoustical Soc Am J	Journal of the Acoustical Society of America (New York)
Acta Biol Med Ger	Acta Biologica et Medica Germanica (Berlin)
Acta Neurol Lat Am	Acta Neurologica Latinoamericana (Montevideo)
Acta Otolaryngol	Acta Oto-Laryngologica (Stockholm)
Acta Otorhinoloryngol Belg	Acta Oto-Rhino-Laryngologica Belgica (Brussels)
Acta Otorinolaryngol Iber Am	Acta Oto-Rino-Laryngologica Ibero-Americana (Barcelona)
Acta Physiol Pol	Acta Physiologica Polonica (Warsaw)
Acta Psychol	Acta Psychologica; European Journal of Psychology (Amsterdam)
Adm Mgt	Administrative Management (New York)
Aeronautical J	Aeronautical Journal (London)
Aerosp Med	Aerospace Medicine (St. Paul)
Air Cond Heat & Refrig N	Air Conditioning Heating and Refrigeration News (Birmingham, Michigan)
Air Pollution Control Assn J	Air Pollution Control Association. Journal (Pittsburgh)
Aircraft Eng	Aircraft Engineering (London)
Am Dyestuff Rep	American Dyestuff Reporter (New York)
Am Fam Physician	American Family Physician (Kansas City, Missouri)
Am Ind Hyg Assoc J	American Industrial Hygiene Association Journal (Detroit)
Am Inst Plan J	American Institute of Planners. Journal. (Washington)

Am J Clin Hypn	American Journal of Clinical Hypnosis (Phoenix)
Am J Men Deficiency	American Journal of Mental Deficiency (Albany)
Am J Obstet Gynecol	American Journal of Obstetrics and Gynecology (St. Louis)
Am J Phys	American Journal of Physics (New York)
Am J Public Helath	American Journal of Public Health and the Nation's Health (New York)
Am J Roentgenol Radium Ther Nuci Med	American Journal of Roentgenology Radium Therapy and Nuclear Medicine (Springfield, Illinois)
Am Paper Ind	American Paper Industry (Des Plaines, Illinois)
Am Sch & Univ	American School and University (New York)
Am Soc C E Proc	American Society of Civil Engineers. Proceedings (New York)
Amer J Physiol	American Journal of Physiology (Bethesda, Maryland)
Amer Med News	American Medical Association News (Chicago)
Ann NY Acad Sci	Annals of the New York Academy of Sciences (New York)
Ann Occup Hyg	Annals of Occupational Hygiene (London)
Ann Otol Rhinol Laryngol	Annals of Otology, Rhinology and Laryngology (St. Louis, Missouri)
Annee Psychol	Annee Psychologique (Paris)
Appraisal J	Appraisal Journal (Chicago)
Arch Belg Med Soc	Archives Belges de Medecine Sociale, Hygiene, Medecine du Travail et Medecine Legale (Brusselles)
Arch Environ Health	Archives of Environmental Health (Chicago)
Arch Int Pharmacodyn Ther	Archives Internationales de Pharmacodynamie et de Therapie (Gand)
Arch Klin Exp Ohren Nasen Kehlkopfheilkd	Archiv fuer Klinische und Experimentelle Ohrennasen- und Kehlkopfheilkunde (Berlin)

Arch Neurobiol	Archivos de Neurobiologia (Madrid)
Arch Otolaryngol	Archives of Otolaryngology (Chicago)
Arch Otorhinolaryngol	Archives of Oto-Rhino-Laryngology (New York)
Arch Phys Med Rehabil	Archives of Physical Medicine and Rehabilitation (Chicago)
Archit Rec	Architectural Record (New York)
Astronautics & Aeronautics	Astronautics & Aeronautics (New York)
Atmospheric Environment	Atmospheric Environment (Elmsford, New York)
Audio	Audio (Philadelphia)
Audio Eng Soc J	Audio Engineering Society. Journal (New York)
Audiology	Audiology (Basel)
Automation	Automation (Cleveland, Ohio)
Automotive Eng	SAE Journal of Automotive Engineering (Warrendale, Philadelphia)
Automotive Ind	Automotive Industries (Philadelphia)
AV Instr	Audiovisual Instruction (Washington)
Aviation W	Aviation Week and Space Technology (New York)
Behav Biol	Behavioral Biology (New York)
Behav Res Ther	Behaviour Research and Therapy (Oxford)
Bell System Tech J	Bell System Technical Journal (Murray Hill, New Jersey)
Bldg Systems Design	Building Systems Design (Brooklyn)
Br Med J	British Medical Journal (London)
Brain Behav Evol	Brain, Behavior and Evolution (Basel)
Brain Res	Brain Research (Amsterdam)
Build Res Estab Digest	Building Research Establishment Digest (London)
Building Oper Manage	Building Operating Management (Milwaukee)
Bull Acad Natl Med	Bulletin de l'Academie Nationale de Medecine (Paris)
Bull NY Acad Med	Bulletin of the New York Academy of Medicine (New York)

Bus W	Business Week (New York)
C R Acad Sci	Comptes rendus hebdomadaires des seances de l'Academie des Sciences; D: Sciences Naturelles (Paris)
Can Bus	Canadian Business (Montreal)
Can J Otolaryngol	Canadian Journal of Otolaryngology (Don Mills, Canada)
Can J Psychol	Canadian Journal of Psychology (Toronto)
Cas Lek Cesk	Casopis Lekaru Ceskych (Prague)
Ceylon Med J	Ceylon Medical Journal (Colombo)
Chem & Eng N	Chemical and Engineering News (Washington)
Chem Eng	Chemical Engineering (New York)
Chem Eng Prog	Chemcial Engineering Progress (New York)
Chem Mktg Rep	Chemical Marketing Report (New York)
Chem W	Chemical Week (New York)
Chemistry	Chemistry (Washington)
Chin Med J	Chinese Medical Journal (Peking)
Civil Eng	Civil Engineering (New York)
Clin Toxicol	Clinical Toxicology (New York)
Combustion	Combustion (New York)
Community Health	Community Health (Bristol)
Comp Air Mag	Compressed Air Magazine (Phillipsburg, New Jersey)
Comput Biomed Res	Computers and Biomedical Research (New York)
Country Life	Country Life (Appleton, Wisconsin)
Daily Telegraph	Daily Telegraph Magazine (London)
Dev Psychobiol	Developmental Psychobiology (New York)
Diesel Equip Supt	Diesel Equipment Superintendent (Stamford, Connecticut)
Dis Nerv Syst	Diseases of the Nervous System (Galveston)
Discussion	Discussion
Dist Nurs	District Nurse (London)
Dtsch Med Wochenschr	Deutsche Medizinische Wochenschrift (Stuttgart)

Dtsch Zahnaerztl Z	Deutsche Zahnaerztliche Zeitschrift (Munich)
Ecology L Q	Ecology Law Quarterly (Berkeley, California)
Economist	Economist (London)
Ed & Pub	Editor & Publisher-the Fourth Estate (New York)
Elec Constr & Maint	Electrical Construction and Maintenance (Highstown, New Jersey)
Elec World	Electrical World (Hightstown, New Jersey)
Electroencephalogr Clin Neurophysiol	Electroencephalography and Clinical Neurophysiology (Amsterdam)
Electronic N	Electronic News (New York)
Electronics & Power	Electronics and Power (Herts, England)
Engin N	Engineering News (London)
Engineer	Engineer (New York)
Engineering	Engineering (London)
Environment	Environment (Bridgeton, Missouri)
Ergonomics	Ergonomics (London)
Eur Neurol	European Neurology (Basel)
Exp Brain Res	Experimental Brain Research (Berlin)
Exp Neurol	Experimental Neurology (New York)
Experientia	Experientia (Basel)
Eye Ear Nose Throat Mon	Eye, Ear, Nose and Throat Monthly (Chicago)
Factory	Factory (Bombay, India)
Fam Hlth	Family Health (New York)
Fed Proc	Federation Proceedings: Federation of American Societies for Experimental Biology (Bethesda)
Fla Envirnomental & Urban Issues	Florida Environmental and Urban Issues (Fort Lauderdale)
Fleet Owner	Fleet Owner (New York)
Folia Morphol	Folia Morphologica (Warsaw)
Folia Phoniatr	Folia Phoniatrica (Basel)
Food Eng	Food Engineering (Philadelphia)
Food Processing	Food Processing (Chicago)
Fortschr Zool	Fortschritte der Zoologie (Stuttgart)
Foundry	Foundry (Cleveland, Ohio)
Fueloil & Oil Heat	Fueloil & Oil Heat (Cedar Grove, New Jersey)

Gesund Ing	Gesundheits-Ingenieur (Munich)
Gig Sanit	Gigiyena i Sanitariya (Moscow)
Gig Tr Prof Zabol	Gigiyena Truda i Professional'nyye Zabolevaniya (Moscow)
Harefuah	Harefuah (Tel Aviv)
Health People	The Health of the People (Auckland, New Zealand)
Health Soc Serv J	Health and Social Service Journal (London)
Health Visit	Health Visitor (London)
Heating-Piping	Heating, Piping & Air Conditioning (Stamford, Connecticut)
Hefte Unfallheilkd	Hefte zur Unfallheilkunde; Beihefte zur Monatsscrift fur Unfallheilkunde, Versicherungs-, Versorgungs und Verkehrsmedizin (Berlin)
Horm Res	Hormone Research (Basel)
Horticulture	Horticulture (Marion, Ohio)
Hosp Eng	Hospital Engineering (Herts, England)
Hosp Top	Hospital Topics (Chicago)
Hot Rod	Hot Rod (Los Angeles)
Hudson R	Husdon Review (New York)
Hydraulics & Pneumatics	Hydraulics and Pneumatics (Cleveland, Ohio)
Hydrocarbon Process	Hydrocarbon Processing-Petroleum Refiner (Houston, Texas)
IEEE Proc	Proceedings of the IEEE (Institute of Electrical and Electronics Engineers) (New York)
IEEE Spectrum	IEEE Spectrum (New York)
IEEE Trans Biomed Eng	IEEE Transactions on Bio-Medical Engineering (New York)
IEEE Trans Com	IEEE Transactions. Communication Technology (New York)
IEEE Trans Ind Applications	IEEE Transactions on Industry Applications (New York)
Impact Sci Soc	Impact of Science on Society (New York)
Ind W	Industrial Worker (Chicago)
Inland Ptr/Am Lith	Inland Printer/American Lithographer (Chicago)

Instr & Control Systems	Instruments and Control Systems (Radnor, Pennsylvania)
Int Arch Arbeitsmed	Internationales Archiv fur Arbeitsmedizin (Berlin)
Internist	Internist (Berlin)
Iron Age	Iron Age (Philadelphia)
J Acoust Soc Am	Journal of the Acoustical Society of America (New York)
J Air L	Journal of Air Law and Commerce (Dallas, Texas)
J Air Pollut Control Assoc	Journal of the Air Pollution Control Association (Pittsburgh)
J Aircraft	Journal of Aircraft (New York)
J Am Dent Assoc	Journal of the American Dental Association (Chicago)
J Am Oil Chem Soc	Journal of the American Oil Chemists' Society (Chicago)
J Am Osteopath Assoc	Journal of the American Osteopathic Association (Chicago)
J Appl Psychol	Journal of Applied Psychology (Washington)
J Arkansas Med Soc	Journal of the Arkansas Medical Society (Fort Smith)
J Commun Disord	Journal of Communication Disorders (Amsterdam)
J Comp Physiol Psychol	Journal of Comparative and Physiological Psychology (Washington)
J Eng Ind	Journal of Engineering for Industry (New York)
J Environmental Health	Journal of Environmental Health (Denver, Colorado)
J Environmental Sci	Journal of Environmental Sciences (Mt. Prospect, Illinois)
J Exp Biol	Journal of Experimental Biology (London)
J Exp Child Psychol	Journal of Experimental Child Psychology (New York)
J Exp Psychol	Journal of Experimental Psychology (Washington)
J Exp Soc Psychol	Journal of Experimental Social Psychology (New York)
J Exp Zool	Journal of Experimental Zoology (Philadelphia)

J Fr Otorhinolaryngol	Journal Francais d'Oto-Rhino-Laryngologie, Audio- Phonologie et Chirurgie Maxillo-Faciale (Lyon)
J Health & Soc Behav	Journal of Health and Social Behavior (Washington)
J Hum Ergol	Journal of Human Ergology (Tokyo)
J Infect Dis	Journal of Infectious Diseases (Chicago)
J Insect Physiol	Journal of Insect Physiology (Oxford)
J La State Med Soc	Journal of the Louisiana State Medical Society (New Orleans)
J Med Assoc State Ala	Journal of the Medical Association of the State of Alabama (Montgomery)
J Ment Defic Res	Journal of Mental Deficiency Research (Caterham)
J Morphol	Journal of Morphology (Philadelphia)
J Neurocytol	Journal of Neurocytology (London)
J Neurophysiol	Journal of Neurophysiology (Bethesda, Maryland)
J Neurosurg	Journal of Neurosurgery (Chicago)
J Occup Med	Journal of Occupational Medicine (Chicago)
J Otolaryngol Jap	Journal of Otolaryngology of Japan (Tokyo)
J Pharm Pharmacol	Journal of Pharmacy and Pharmacology (London)
J Phys E. Sci Instrum	Journal of Physics E: Scientific Instruments (London)
J Physiol	Journal de Physiologie (Paris)
J Physiol Soc Jap	Journal of the Physiological Society of Japan (Tokyo)
J Psycholinguist Res	Journal of Psycholinguistic Research (New York)
J Reprod Fertil	Journal of Reproduction and Fertility (Oxford)
J Sch Health	Journal of School Health (Columbus)
J Speech & Hearing Dis	Journal of Speech and Hearing Disorders (Washington)
J Speech & Hearing Res	Journal of Speech and Hearing Research (Washington)

Jap J Clin Med	Japanese Journal of Clinical Medicine (Osaka)
Jap J Hyg	Japanese Journal of Hygiene (Kyoto, Japan)
Jap J Med Electron	Japanese Journal of Medical Electronics and Biological Engineering (Tokyo)
Kan L. Rev	University of Kansas Law Review (Lawrence)
Klin Med	Klinicheskaia Meditsina (Moscow)
Klin Paediatr	Klinische Paediatrie (Stuttgart)
Lab Anim Sci	Laboratory Animal Science (Joliet, Illinois)
Labor Research	Labor Research (London)
Labour Gaz	Labour Gazette (Ottawa)
Ladies Home J	Ladies Home Journal (New York)
Lakartidningen	Lakartidningen (Stockholm)
Lancet	Lancet (London)
Laryngol Rhinol Otol	Laryngologie, Rhinologie, Otologie und Ihre Grenzgebiete (Stuttgart)
Laryngoscope	Laryngoscope (Collinsville, Illinois)
Life Health	Life and Health (Washington)
Machine Design	Machine Design (Cleveland, Ohio)
Maine L Rev	Maine Law Review (Portland)
Man/Soc/Tech	Man/Society/Technology (Washington)
Manuf Eng & Mgt	Manufacturing Engineering and Management (Dearborn, Michigan)
Materials Eng	Materials Engineering (Stamford, Connecticut)
Mech Eng	Mechanical Engineering (New York)
Med Lav	Medicina de Lavaro (Milano)
Med Radiol	Meditsinskaia Radiologiia (Moscow)
Med Times	Medical Times (Manhasset)
Mel Maker	Melody Maker (London)
Metallurgia & Metal Forming	Metallurgia and Metal Forming (London)
Mgt R	Management Review (New York)
Mich Hosp	Michigan Hospitals (Lansing)
Min Cong J	Mining Congress Journal (Washington)
Minn L Rev	Minnesota Law Review (Minneapolis)

Mod Health Care	Modern Health Care (New York)
Mod Materials Handling	Modern Materials Handling (New York)
Motor B	Motor Business (London)
Mus J	Music Journal (New York)
N Engl J Med	New England Journal of Medicine (Boston)
Nar Zdrav	Narodno Zdravlje (Belgrade)
Natural Resources Law	Natural Resources Lawyer (Chicago)
Nature	Nature (London)
Neuropsychologia	Neuropsychologia (Elmsford, New York)
Neuroradiology	Neuroradiology (Berlin)
Newsweek	Newsweek (New York)
Nor Tannlaegeforen Tid	Norske Tannlaegeforenings Tidende (Oslo)
Nouv Presse Med	Nouvelle Presse Medicale (Paris)
Nurs Times	Nursing Times (London)
Nursing	Nursing (Jenkintown)
Nursing Care	Nursing Care (New York)
NY L F	New York Law Forum (New York)
NYU Res L & Soc Change	New York University Research Law and Social Change
ORL	ORL; Journal for Oto-Rhino-Laryngology and its Borderlands (Basel)
Occup Health	Occupational Health (London)
Ohio State Med J	Ohio State Medical Journal (Columbus)
Oil & Gas J	Oil and Gas Journal (Tulsa, Oklahoma)
Okla L Rev	Oklahoma Law Review (Norman)
Optimum	Optimum (Ottawa)
OR Reporter	OR Reporter (Southfield, Massachusetts)
Otolaryngol Pol	Otolaryngologia Polska (Warsaw)
Pediatrics	Pediatrics (Springfield, Illinois)
Percept Motor Skills	Perceptual and Motor Skills (Missoula, Montana)
Perception	Perception (London)
Pers J	Personnel Journal (Santa Monica)
Pfluegers Arch	Pfluegers Archiv; European Journal of Physiology (Berlin)

Phys Teach	Physics Teacher (Stonybrook, New York)
Physiol Behav	Physiology and Behavior (Oxford)
Pieleg Polozna	Pielegniarka i Polozna (Warsaw)
Pipeline & Gas J	Pipeline and Gas Journal (Dallas)
Pit & Quarry	Pit and Quarry (Chicago)
Plant Eng	Plant Engineering (Barrington, Illinois)
Plastics Eng	Plastics Engineering (Greenwich, Connecticut)
Pol Przegl Radiol	Polski Przeglad Radiologii i Medycyny Nuklearnej (Warsaw)
Pol Tyg Lek	Polski Tygodnik Lekarski (Warsaw)
Pop Electr	Popular Electronics (New York)
Pol Mech	Popular Mechanics (New York)
Power	Power (New York)
Power Eng	Power Engineering (Barrington, Illinois)
Probl Actuels Otorhinolaryngol	Problemes Actuels d'Oto-Rhino-Laryngologie (Paris)
Proc Natl Acad Sci USA	Proceedings of the National Academy of Sciences of the United States of America (Washington)
Product Eng	Product Engineering (New York)
Psychiatr Neurol Med Psychol	Psychiatrie, Neurologie und Medizinische Psychologie (Leipzig)
Psychol Med	Psychological Medicine (London)
Psychol Rep	Psychological Reports (Missoula, Montana)
Psychophysiology	Psychophysiology (Detroit, Michigan)
Pub Works	Public Works (Ridgewood, New Jersey)
Pulp & Pa	Pulp & Paper (New York)
Purchasing	Purchasing (New York)
Q J Exp Physiol	Quarterly Journal of Experimental Physiology and Cognate Medical Sciences (Edinburgh)
Q J Exp Psychol	Quarterly Journal of Experimental Psychology (Cambridge, England)
Quality Prog	Quality Progress (Milwaukee)
R Soc Health J	Royal Society of Health Journal (London)

Radio-Electr	Radio-Electronics (New York)
Radio Sci	Radio Science (Washington)
Read Digest	Readers Digest (Pleasantville, New York)
Rev Med Chil	Revista Medica de Chile (Santiago)
Rev Med Chir Soc Med Nat IASl	Revista Medica-Chirurgicala a Societatie di Medici si Naturalisti din Iasi
Rev Med Suisse Romande	Revue Medicale de al Suisse Romande (Lausanne)
Rev Roum Neurol	Revue Roumaine de Neurologie (Bucharest)
Rev Sci Instrum	Review of Scientific Instruments (New York)
Roads & Sts	Roads and Streets (New York)
Roentgenblaetter	Roentgenblaetter (Wuppertal)
Roy Town Plan Inst J	Royal Town Planning Institute Journal (London)
Rubber World	Rubber World (New York)
Rutgers-Camden L J	Rutgers-Camden Law Journal (Camden)
SMPTE J	Society of Motion Picture and Television Engineers (Scarsdale, New York)
Sb Ved Pr Lek Fak Karlovy Univ	Sbornik Vedeckych Praci Lekarski Fakulty Karlovy University (Hradec, Kralove)
Sch Mus	School Musician Director and Teacher (Joliet, Illinois)
Sci & Child	Science and Children (Washington)
Sci Digest	Science Digest (New York)
Sci Instr	Scientific Instruments (Journal of Physics E.) (London)
Science	Science (Washington)
Sist Nerv	Sistema Nervoso (Milan)
So Calif L Rev	Southern California Law Review (Los Angeles, California)
Society	Society (New Brunswick, New Jersey)
Sov J Dev Biol	Soviet Journal of Developmental Biology (New York); English Edition of Ontogenez (Moscow)

Special Libraries	Special Libraries (New York)
Superv Nurse	Supervisor Nurse (Chicago)
Surg Neurol	Surgical Neurology (Tryon, North Carolina)
Sygeplejersken	Sygeplejersken (Copenhagen)
Syracuse L Rev	Syracuse Law Review (Syracuse)
TAPPI	Technical Association of the Pulp and Paper Industry (Atlanta, Georgia)
Tex Bus R	Texas Business Review (Austin)
Tex Int L J	Texas International Law Journal (Austin)
Textile World	Textile World (New York)
Ther Hung	Therapia Hungarica (Budapest)
Tidsskr Nor Laegeforen	Tidsskrift for den Norske Laegeforening (Oslo)
Tohoku J Exp Med	Tohoku Journal of Experimental Medicine (Sendai)
Traffic Q	Traffic Quarterly (Saugatuck, Connecticut)
Trans Am Acad Ophthalmol Otolaryngol	Transactions of the American Academy of Ophthalmology and Otolaryngology (Rochester, Minneapolis)
U S News	U S News and World Report (Washington)
Urban L Ann	Urban Law Annual (St. Louis, Missouri)
Urol Int	Urologia Internationalis (Basel)
Vestn Akad Med Nauk SSSR	Vestnik Akademii Meditsinskikh Nauk SSSR (Moscow)
Vestn Dermatol Venerol	Vestnik Dermatologii i Venerologii (Moscow)
Vestn Oftalmol	Vestnik Oftalmologii (Moscow)
Vestn Otorinolaringol	Vestnik Otorinolaringologii (Moscow)
Vision Res	Vision Research (Oxford)
Voen Med Zh	Voenno-Meditsinskii Zhurnal (Moscow)
Vopr Neirokhir	Voprosy Neirokhirurgii (Moscow)
Vrach Delo	Vrachebnoe Delo (Kiev)
W Va Med J	West Virginia Medical Journal (Charleston)

Wall St J	Wall Street Journal (New York)
Washburn L J	Washburn Law Journal (Topeka, Kansas)
Willamette L J	Willamette Law Journal (Salem, Oregon)
Z Exp Chir	Zeitschrift fur Experimentelle Chirurgie (Berlin)
Z Gesamte Hyg	Zeitschrift fur die Gesamte Hygiene und Ihre Grenzgebiete (Berlin)
Z Gesamte Inn Med	Zeitschrift fur die Gesamte Innere Medizin und Ihre Grenzgebiete (Leipzig)
Z Kardiol	Zeitschrift fur Kardiologie (Darmstadt)
Z Laryngol Rhinol Otol	Zeitschrift fuer Laryngologie, Rhinologie, Otologie und Ihre Grenzgebiete (Stuttgart)
Z Urol Nephrol	Zeitschrift fur Urologie und Nephrologie (Leipzig)
Zentralbl Arbeitsmed	Zentralblatt fur Arbeitsmedizin und Arbeitsschutz (Darmstadt)
Zh Obshch Biol	Zhurnal Obshchei Biologii (Moscow)
Zh Ushn Nos Gorl Bolezn	Zhurnal Ushnykh, Nosovyky i Gorlovykh Boleznei (Kiev)

SUBJECT HEADINGS USED IN THIS BIBLIOGRAPHY

Acoustic Nerve
Acoustic Stimulation
Acoustical Society of America
Air Conditioning
Aircraft Noise
Airplanes
Airplanes: Model
Airports
Amplitude Discrimination
Angiotensin
Architecture
Automobile Noise
Barotrauma
Behavior
Bekesy Typing
Boilers
Breath
Bus Noise
Caffeine Citrate
Cardiovascular System
Cataracts
Children
Churches
Circulatory System
Clindamycin
Clinical Aspects
Combustion Noise
Community Noise
Compressed Air Noise
Compressor Noise
Dental Noise
Diabetes
Diazepam
Discotheques
Drugs
Earmuffs
Earphones
Earplugs
Economic Aspects
Electronic Noise

Employees
Engines: Diesel
Engines: Jet
Environmental Health
EPA
Ether
FAA
Fans: Electric
Fans: Mechanical
Fireworks
Fishing Vessels
Furnace Noise
Hearing Aids
Hearing Conservation
Hearing Measurement
Hearing Standards
Helicopter Noise
Hospital Noise
Household Noise
Hunters
Hygienic Studies
Impact Noise
Impulse Noise
Industrial Hygiene
Industrial Medicine
Industrial Noise
Industrial Noise: Aeroplane Factory
Industrial Noise: Construction
Industrial Noise: Flour
Industrial Noise: Glass
Industrial Noise: Nail Factory
Industrial Noise: Ore Dressiny Factories
Industrial Noise: Rubber
Industrial Noise: Shoe-Factory
Industrial Noise: Textile
Inner Ear
Instrumentation
Iophendylate

Kanamycin
Laboratories
Landscaping
Larynx
Laws and Legislation
Learning
Libraries
Lightning
Machine Design
Machinery Noise
Machinery Noise: Electric
Machinery Noise: Foundry
Machinery Noise: Hydraulic
Machinery Noise: Lathes
Machinery Noise: Paper Making
Machinery Noise: Textile
Man
Masking Noise
Metallic Mercury
Methionin-S35
Microphones
Middle Ear
Minibikes
Mining Noise
Motor Trucks
Motors: Electric
Muscles: Skeletal
Music
Musical Instruments
Musicians
NASA
Neomycin
Neonatal
Nervous System
Nitrous Oxide
Noise: Austria
Noise: Canada
Noise: Germany
Noise: Italy
Noise: Japan
Noise: Mexico
Noise: Poland

Noise: Sweden
Noise: United Kingdom
Noise: United States
    California
    Connecticut
    Florida
    Hawaii
    Illinois
    Indiana
    Louisiana
    Massachusetts
    New York
    Oregon
    Wisconsin
    Washington
Noise: USSR
Noise Abatement
Noise Levels
Noise Measurement
Noise Measurement Devices
Noise Research
Noise Standards
Noise Studies
Nortriptyline Hydrochloride
Occupational Deafness
Occupational Health
Oceans
Office Noise
Office Noise: Printing
Otolaryngolgoy
Otology
Otorhinolaryngology
Ototoxicoses
Perception
Performance
Pharmacies
Physicians
Physiology
Pipelines
Plants
Plants: Automobile
Plants: Cellulose-Paper

Plants: Chemical
Plants: Cleaning
Pneumatic Machinery & Tools
Pressure Boxes
Protection
Psilocybine
Psychology
Psychotics
Radio Noise
Railway Noise
Recreational
Respiratory System
Schools
Sleep
Small Borearms
Snowmobiles
Sociology
Sodium Potassium
Sodium Salicylate
Sonic Boom
Soundproofing
Speech
SST
Statistics
Subways
Surgery
Symposia
Telephones
Temporary Threshold Shift
Thunder
Tinnitus
Tractor Operators
Tractors
Traffic Noise
Transmissions
Trivastal
Turbines
Turbines: Gas
Urban Noise
Urban Planning
Valve Noise
Vibration

Vision
Weavers
White Noise
Wind
Workshops
Youth

TABLE OF CONTENTS

Preface i

List of Periodical Abbreviations iii

Subject Headings Used in this Bibliography xvii

Books 1

Periodical Literature:

    Title Index. 5

    Subject Index. 70

    Author Index 168

BOOKS

Alexandre, Ariel, et al. LE TEMPS DU BRUIT. Paris: Flammarion, 1973.

Australian Acoustical Society. NOISE LEGISLATION AND REGULATION: the proceedings of the 1972 annual conference of the Australian Acoustical Society held at the Hotel Florida, Terrigal, N.S.W., 29th September to 2d October, 1972. Kew, Victoria: Australian Acoustical Society, 1972.

Aylesworth, Thomas G. THIS VITAL AIR, THIS VITAL WATER: man's environmental crisis. Chicago: Rand McNally, 1973.

Bartholomew, Robert. THE SONIC ENVIRONMENT AND HUMAN BEHAVIOR. (Exchange bibliography no. 565). n. p.: Council of Planning Librarians, 1974.

Beck, Gerhard. PFLANZEN ALS MITTEL ZUR LARMBEKAMPFUNG. Hanover, Berlin: Sarstedt, Patzer, 1967.

Burns, William. NOISE AND MAN. 2nd ed. London: J. Murray; Philadelphia: Lippincott, 1973.

Conference on Noise and Vibration Control. PROCEEDINGS. Cardiff: University of Wales. Inst. of Science and Technology, 1974.

Conley, Vesta B., et al. DAYTIME NOISE ENVIRONMENT IN TUCSON, ARIZONA. Tucson: Engin, 1973.

Cremer, Lothar, et al. STRUCTURE-BORNE SOUND; structural vibrations and sound radiation at audio frequencies. Translated and reviewed by E. E. Ungar. Berlin, New York: Springer-Verlag, 1973.

Everest, F. A. ACOUSTIC TECHNIQUES FOR HOME AND STUDIO. Blue Ridge Summit, Pennsylvania: TAB Books, 1973.

Fath, Jack M. STANDARDS ON NOISE MEASUREMENTS, rating schemes and definitions. Washington: G.P.O., 1973.

Floyd, Mary K. A BIBLIOGRAPHY OF NOISE FOR 1972. Troy, New York: Whitston Publishing Company, 1974.

Harris, R. W., et al. INTRODUCTION TO NOISE ANALYSIS. New York: Academic Press, n.d.

International Congress on Noise as a Public Health Problem, Dubrovnik, Yugoslavia, 1973. PROCEEDINGS. Washington: G.P.O., 1974.

International Conference on Noise Control Engineering, Washington, 1972. INTER-NOISE 72; proceedings. Edited by Malcolm J. Crocker. Poughkeepsie, New York, 1972.

Isolation of mechanical vibration, impact, and noise; A COLLOQUIUM PRESENTED AT THE ASME DESIGN ENGINEERING TECHNICAL CONFERENCE, CINCINNATI, OHIO, SEPTEMBER, 1973. Edited by John C. Snowdon, et al. New York: American Society of Mechanical Engineers, 1973.

Liebig, Joachim, et al. LARMSCHUTZ BEI INVESTITIONEN. Vertragsrechtl. Probleme u. schalltechn. Grundbegriffe f. d. Larmschutz. Berlin: Verl. Tribune, 1972.

Lipscomb, D. M. NOISE: the unwanted sounds. Chicago, Illinois: Nelson-Hall, 1974.

Lyon, R. H. LECTURES IN TRANSPORTATION NOISE. Harvard, Massachusetts: Grozier Publishing, 1973.

Maschinenlarm auf Baustellen; LEITFADEN ZUR MESSUNG, Beurteilung und Planung. Wiesbaden, Bauverlag, 1973.

Petrusewicz, S. A., et al. NOISE AND VIBRATION CONTROL FOR INDUSTRIALISTS. New York: American Elsevier Publishing, Company, 1974.

Purdue Noise Control Conference, 1971. NOISE AND VIBRATION CON-

CONTROL ENGINEERING; proceedings. Edited by Malcolm J. Crocker. Lafayette, Indiana: Purdue University, 1972.

Sataloff, Joseph, et al. HEARING CONSERVATION. Springfield, Illinois: Thomas, 1973.

Schafer, R. THE BOOK OF NOISE. Vancouver: Price Pint, 1970.

Stafford, James Raymond, et al. DORMITORY NOISE ABATEMENT. Morgantown: Engineering Experiment Station, West Virginia University, 1972.

Stephens, R. W. B., et al. ACOUSTICS AND VIRBATION PROGRESS. Vol. 1. New York: Halsted Press; London: Chapman & Hall, 1974.

Symposium on Performance under Sub-Optimal Conditions, London, 1970. PROCEEDINGS. London: Taylor and Francis, 1970.

Tatusesco, Dan, et al. PROTECTION ACOUSTIQUE DES LOGEMENTS; application de l'arrete du 14 juin 1969. Etude effectuee sous la direction du RAUC. Paris: Eyrolles, 1974.

United States. Congress. Committee on science and astronautics. Subcommittee on aeronautics and space technology. Washington: G.P.O., 1974.

United States. Congress. Senate. Committee on Commerce. Aviation Subcommittee. OVERSIGHT HEARING ON NOISE CONTROL ACT. Hearing before the Subcommittee on Aviation of the Committee on Commerce, United States Senate, Ninety-third Congress, first session, March 30, 1973. Washington: G.P.O., 1973

United States. Congress. Senate. Committee on Public Works. IMPLEMENTATION ON THE NOISE CONTROL ACT OF 1972 (aircraft-airport noise): hearing, pt. 1, September 24, 1973 (93rd Congress, 1st session); pt. 2, March 22, 1974 (93rd Congress, 2nd session). (Serial no. 93-H31.) Washington: G.P.O., 1974.

United States. Congress. Senate. Committee on Public Works. Subcommittee on air and water pollution. REPORT ON AIRCRAFT-AIRPORT NOISE: report of the administrator of the Environmental protection agency. Washington: G.P.O., 1973.

United States. Office of Noise Abatement. REPORT NO. DOT—TST. Washington: G.P.O., 1974

Wiggins, John Henry. EFFECTS OF SONIC BOOM. Redondo Beach, California: J. H. Wiggins, Co., 1969.

PERIODICAL LITERATURE

TITLE INDEX

"ABC's of sound reinforcement," by M. Koller. RADIO-ELECTRONICS 45:40, August, 1974.

"Acoustic and vestibular defects in lightning survivors," by J. W. Wright, Jr., et al. LARYNGOSCOPE 84(8):1378-1387, August, 1974.

"Acoustic conduction of the auditory ossicles," by E. I. Volpliushkin, et al. ZH USHN NOS GORL BOLEZN 33:12-15, 1973.

"Acoustic jaw reflex in man: its relationship to other brain-stem and microreflexes," by K. Meier-Ewert, et al. ELECTROENCEPHALOGR CLIN NEUROPHYSIOL 36:629-637, June, 1974.

"Acoustic neurinoma. A comparison of the clinical picture and the electroencephalogram," by Z. Mensikova, et al. SB VED PR LEK FAK KARLOVY UNIV 15:401-408, 1972.

"Acoustic neurinoma presenting as subarachnoid hemorrhage. Case report," by K. McCoyd, et al. J NEUROSURG 41(3):391-393, September, 1974.

"Acoustic neurinomas presenting as middle ear tumors," by L. A. Storrs. LARYNGOSCOPE 84:1175-1180, July, 1974.

"Acoustic neuroma in the last months of pregnancy," by J. Allen, et al. AM J OBSTET GYNECOL 119:516-520, June 15, 1974.

"The acoustic reflex in eighth nerve disorders," by J. Jerger, et al. ARCH OTOLARYNGOL 99:409-413, June, 1974.

"Acoustic trauma after double exposure in mammals," by A. Pye. AUDIOLOGY 13(4):320-325, 1974.

"Acoustical building panels quieten turbine compressor stations." PIPELINE & GAS J 201:42 plus, May, 1974.

"The acoustical engineer's viewpoint of hearing aid design," by S. F. Lybarger. BULL NY ACAD MED 50:917-930, September, 1974.

"Acoustical floor covering comes of age." AM SCH & UNIV 46,3:35-36, November, 1973.

"Acoustical privacy in the landscaped office," by A. C. C. Warnock. ACOUSTICAL SOC AM J 53:1535-1543, June, 1973.

"Acoustical value of carpeting," by D. B. Parlin. BUILDING OPER MANAGE 21:48 plus, September, 1974.

"Acoustically evoked potentials (a.e.p.) in neonates with special consideration in intrauterine dystrophia," by G. Muller, et al. KLIN PAEDIATR 185:449-457, November, 1973.

"Acoustics in air conditioning." HOSP ENG 27:8-19, January, 1973.

"Acoustics of educational facilities," by E. P. Caffarella, Jr. AV INSTR 18:10-11, December, 1973.

"Action of noise on oxygen consumption in different brain structures," by S. V. Alekseev, et al. GIG SANIT 38:110-111, July, 1973.

"Active machine tool controller requirements for noise attenuation," by E. E. Mitchell, et al. J ENG IND 96:261-267, February, 1974.

"Acute acoustic trauma in a locomotive engineer," by I. E. Zaslavskii. ZH USHN NOS GORL BOLEZN 0(1):112-113, January-February, 1974.

"Acute effects of ethanol on spontaneous and auditory evoked electrical activity in cat brain," by R. G. Perrin, et al. ELECTROENCEPHALOGR CLIN NEUROPHYSIOL 36:19-31, January, 1974.

"Adaptation for sound localization in the ear and brainstem of mammals," by R. B. Masterton. FED PROC 33:1904-1910, August, 1974.

"Adrenal insufficiency and electrophysiological measure of auditory sensitivity," by F. W. Conn, et al. AM J PHYSIOL 225:1430-1436, December, 1973.

"Advantages of quasi-simultaneous stimulation in ERA," by D. Krell, et al. AUDIOLOGY 13(4):342-348, 1974.

"Aircraft environmental problems," by V. L. Blumenthal, et al. J AIRCRAFT 10:529-537, September, 1973.

"Aircraft noise abatement via annex 16 of the Chicago convention—a viable alternative," by S. S. Kalsi. TEX INT L J 9:1-18, Winter, 1974.

"Aircraft noise and psychiatric morbidity," by F. Gattoni, et al. PHYCHOL MED 3:516-520, November, 1973.

"Aircraft noise induced vibration in fifteen residences near Seattle-Tacoma International Airport," by S. M. Cant, et al. AM IND HYG ASSOC J 34:463-468, October, 1973.

"Aircraft noise now on-line to Stockholm's new pollution monitoring network." ATMOSPHERIC ENVIRONMENT 7:Suppl:2-3, December, 1973.

"Airport noise unaffected by flight cuts," by W. A. Shumman. AVIATION W 100:29-30, February 4, 1974.

"Airport officials hit FAA, DOT on noise, fund issues," by E. J. Bulban. AVIATION W 99:34 plus, October 29, 1973.

"Alterations in morphine-induced analgesia in mice exposed to pain, light or sound," by M. W. Stevens, et al. ARCH INT PHARMACODYN THER 206:66-75, November, 1973.

"Analysis of acoustic signal registered during respirofonomerty," by I. Simacek. CAS LEK CESK 113(37):1122-1124, September 13, 1974.

"Analysis of central nervous system involvement in the microwave auditory effect," by E. M. Taylor, et al. BRAIN RES 74:201-208, July 12, 1974.

"Analysis of the degree of influence of environmental factors with multiple combined action," by R. E. Sova, et al. GIG TR PROF ZABOL 18:46-48, February, 1974.

"Analysis of errors in measuring machine noise under free-field conditions," by G. Hubner. ACOUSTICAL SOC AM J 54:967-977, October, 1973.

"Analysis of evoked responses in man elicited by sinusoidally modulated noise," by M. Rodenburg, et al. AUDIOLOGY 11:283-293, September-December, 1972.

"Analysis of information-bearing elements in complex sounds by auditory neurons of bats," by N. Suga. AUDIOLOGY 11:58-72, January-April, 1972.

"An analysis of sensory interaction," by R. L. Taylor. NEUROPSYCHOLOGIA 12:65-71, January, 1974.

"Analyzing the sounds of trouble," by R. E. Herzog. MACHINE DESIGN 45:128-134, September 6, 1973.

"Angiographic diagnosis of acoustic neurinomas: analysis of 30 lesions," by M. Takahashi, et al. NEURORADIOLOGY 2:191-200, September, 1971.

"Animal and human tolerance of high-dose intramuscular therapy with spectinomycin," by E. Novak, et al. J INFECT DIS 130:50-55, July, 1974.

"Annoyance judgments of aircraft with and without acoustically treated nacelles," by P. N. Borsky. J ACOUST SOC AM 55:Suppl:67, 1974.

"Another way to measure noise [dosimeters]." FACTORY 7:55, February, 1974.

"Antinoise ear plug made of film porolon," by V. Ia. Gapanovich, et al. GIG TR PROF ZABOL 16:54, July, 1972.

"Apparatus for studying the hearing function by means of impedance measuring," by B. S. Moroz, et al. ZH USHN NOS GORL BOLEZN 0(1): 115–117, January-February, 1974.

"Application of constrained-layer damping to control noise in machine parts," by P. D. Emerson. J ENG IND 96:299-303, February, 1973.

"Application of some new survey techniques for assessing exposure to noise and human reaction," by M. Braden, et al. J ACOUST SOC AM 55(2): 464, 1974.

"Arousal and recall: effects of noise on two retrieval strategies," by S. Schwartz. J EXP PSYCHOL 102:896-898, May, 1974.

"Arousal from sleep: the differential effect of frequencies equated for loudness," by T. E. Levere, et al. PHYSIOL BEHAV 12:573-582, April,

"Articulatory interpretation of the 'singing formant'," by J. Sundberg. J ACOUST SOC AM 55:838-844, April, 1974.

"Assessment of acoustic trauma," (proceedings), by F. Schwetz, et al. HEFTE UNFALLHEILKD 114:197-220, 1973.

"The assessment of occupational noise exposure," by A. M. Martin. ANN OCCUP HYG 16(4):353-362, 1973.

"Attempt at physical characterization of the passive sound behavior in the lung on a model," by H. R. Bohme. Z GESAMTE INN MED 29:401-406, May 15, 1974.

"Attenuation characteristics of recreational helmets," by F. H. Bess, et al. ANN OTOL RHINOL LARYNGOL 83:119-124, January-February, 1974.

"Audible noise levels of oxygen masks operating on venturi principle," by J. M. Leigh. BR MED J 4:652, December 15, 1973.

"Audiologic evaluation in cochlear and eighth nerve disorders," by J. W. Sanders, et al. ARCH OTOLARYNGOL 100(4):283-289, October, 1974.

"Audiological comparison of cochlear and eighth nerve disorders," by J. Jerger, et al. ANN OTOL RHINOL LARYNGOL 83:275-285, May-June, 1974.

"Audiometric and anatomical correlates of impulse noise exposure," by D. Henderson, et al. ARCH OTOLARYNGOL 99:62-66, January, 1974.

"Auditory electromyographic feedback therapy to inhibit undesired motor activity," by D. Swaan, et al. ARCH PHYS MED REHABIL 55:251-254, June, 1974.

"Auditory evoked potentials: developmental changes of threshold and amplitude following early acoustic trauma," by J. F. Willott, et al. J COMP PHYSIOL PSYCHOL 86:1-7, January, 1974.

"Auditory information processing of vocalization," by K. Murata, et al. J PHYSIOL SOC JAP 35:535, August-September, 1973.

"The auditory stimuli to evoke a clear average response at behavioral threshold," by S. Funasaka, et al. AUDIOLOGY 13:162-172, March-April, 1974.

"Automatic sound-intensity amplifiers-storage in electric hearing aids," by F. J. Landwehr, et al. ARCH KLIN EXP OHREN NASEN KEHIKOPF-HEILKD 205:252-256, December 17, 1973.

"Automatic urban noise monitoring and analysis system," by J. E. K. Foreman, et al. ACOUSTICAL SOC AM J 55:1358-1359, June, 1974.

"Averaged electroencephalographic response to intensity modulated tone," by T. Okitsu. TOHOKU J EXP MED 112:315-323, April, 1974.

"Azure B-RNA changes in the adrenal and cerebral cortex of rats exposed to intense noise," by A. Anthony. FED PROC 32:2093-2097, November, 1973.

"Background noise study in Chicago," by C. Caccavari, et al. AIR POLLUTION CONTROL ASSN J 24:240-244, March, 1974.

"A backward glance at noise pollution," by G. Rosen. AM J PUBLIC HEALTH 64:514-517, May, 1974.

"Barriers for noise control," by J. N. Macduff. MECH ENG 96:26-31, August, 1974.

"Barth's myochordotonal organ as an acoustic sensor in the ghost crab,

Ocypode," by K. Horch. EXPERIENTIA 30:630-631, June 15, 1974.

"Bayesian density functions for Gaussian pulse shapes in Gaussian noise," by R. Clow, et al. IEEE PROC 62:134-136, January, 1974.

"Benefit for hearing loss." OCCUP HEALTH 25:449-450, December, 1973.

"Better method for air system acoustical design," by J. A. Reese, et al. ASHRAE J 16:59-63, September, 1974.

"Big noise for Illinois," by F. M. H. Gregory. HOT ROD 27:54-55, January, 1974.

"Big noises are being heard as industry considers the cost," by C. Beatson. ENGINEER 238:38-39 plus, February 7, 1974.

"Biological modeling and criteria for standardization of whole-body vibration and noise," by E. Ts. Andreeva-Galanina, et al. VESTN AKAD MED NAUK SSSR 28:30-37, 1973.

"Biopotentials of the organ of hearing in chronic sodium fluoride poisoning," by M. Kowalewaska. OTOLARYNGOL POL 28(4):417-424, 1974.

"Bituminous overlay reduces traffic noise," by H. O. Klossner. PUB WORKS 104:82-83, July, 1973.

"Blasting noise research project." PIT & QUARRY 66:26-27, March, 1974.

"Body movements in sleep during 30-day exposure to tone pulse," by A. G. Muzet, et al. PSYCHOPHYSIOLOGY 11:27-34, January, 1974.

"Breath-sound changes after cigarette smoking," by C. W. Laird, et al. LANCET 1:808, April 27, 1974.

"Brief-tone audiometry with normal, cochlear, and eighth nerve tumor patients," by W. O. Olsen, et al. ARCH OTOLARYNGOL 99:185-189, March, 1974.

"Bring noise to book." ENGINEER 237:23, September 20, 1973.

"Build-them-yourself kits to muffle machine noise." ENGINEER 238:23

April 11, 1974.

"Building noise in a hospital: an experimental simulation," by M. Powell. ANN OCCUP HYG 16:77-79, April, 1973.

"Burwen dynamic noise filter," by B. Whyte. AUDIO 58:10 plus, February, 1974.

"Business flying faces new challenges; Hawker Siddeley, Rolls to test HS.125 with sound suppression," by H. J. Coleman. AVIATION W 99:57-58, September 24, 1973.

"Calculating an equivalent level of unstable noise," by E. I. Denisov. GIG TR PROF ZABOL 17(7):50-51, 1973.

"Calculating noise levels of fans; nomograph," by F. Caplan. PLANT ENG 28:88-89, January 24, 1974.

"Calculation of an equivalent level of nonstable noise," by E. I. Denisov, et al. GIG TR PROF ZABOL 17:50-51, July, 1973.

"California court bars class action; suit to recover damages for aircraft noise." AVIATION W 101:35, October 7, 1974.

"Can we achieve a quieter environment?" by K. S. Oliphant. ASTM STAND N 2:8-13, May, 1974.

"Carbide plant too noise?" CHEM W 114:15, March 13, 1974.

"Cerebral vascular accidents in the course of tumors of the cerebellopontine angle. Pathogenic considerations," by C. Arseni, et al. EUR NEUROL 10:144-159, 1973.

"Certification of Concorde, Tu-144 backed by environmental unit." AVIATION W 101:297, July 15, 1974.

"Change in heart rate due to acoustic stimulation, audiologic test-method," by H. Chuden. ARCH KLIN EXP OHREN NASEN KEHIKOPF-HEILKD 205:231-238. December 17, 1973.

"Change in the reactivity of the terminal vessels of the rat brain under the action of stable noise," by S. V. Alekseev, et al. GIG TR PROF ZABOL

17:18-21, July, 1973.

"Changes of heart rate associated with responses to cyclic visual and acoustic stimuli," by K. Scheuch, et al. ACTA BIOL MED GER 32(4):385-391, 1974.

"Changes of hippocampal single-cell activity in emotion and motivation-active stimuli," by U. Zippel, et al. ACTA BIOL MED GER 31:841-851, 1973.

"Characteristics of the cardiovascular system of adolescent workers subjected to the action of stable industrial noise," by E. A. Gel'tishcheva. GIG TR PROF ZABOL 16:29-33, July, 1973.

"Chronic intracochlear electrode implantation: cochlear pathology and acoustic nerve survival," by R. A. Schindler, et al. ANN OTOL RHINOL LARYNGOL 83:202-215, March-April, 1974.

"City of Burbank v. Lockheed Air Terminal, Inc. (93 Sup Ct 1854): federal preemption of aircraft noise regulation and the future of proprietary restrictions." NYU REV L & SOC CHANGE 4:99-113, Winter, 1974.

"Cochlear electrical activity in noise-induced hearing loss, behavioral and electrophysiological studies in primates," by J. E. Pugh, Jr., et al. ARCH OTOLARYNGOL 100:36-40, July, 1974.

"Cochlear findings in 8th nerve tumors," by S. Katinsky, et al. AUDIOLOGY 11:213-217, May-August, 1972.

"Cochlear neurons: frequency selectivity altered by perilymph removal," by D. Robertson. SCIENCE 186:153-155, October 11, 1974.

"Cochlear pathology in monkeys exposed to impulse noise," by V. M. Jordan, et al. ACTA OTOLARYNGOL 16-30, 1973.

"Cochleo vestibular disturbances in vibration disease," by M. A. Nekhorosheva, et al. GIG TR PROF ZABOL 17(7):9-11, 1973.

"The co-existence of acoustic neuroma and otosclerosis," by J. D. Clemis. LARYNGOSCOPE 83:1959-1985, December, 1973.

"Combination effects of carbon tetrachloride and noise on GPT and LAP activity in serum of rats," by G. Wagner, et al. Z GESAMTE HYG 19:862-863, 1973.

"Combined action of ultrasonics and noise of standard parameters," by A. V. Il'nitskaia, et al. GIG SANIT 38:50-53, May, 1973.

"Combustion noise," by F. E. J. Briffa, et al. COMBUSTION 45:27-37, March, 1974.

"Committee on Environmental Hazards. Noise pollution: neonatal aspects," PEDIATRICS 54(4):476-479, October, 1974.

"Community noise control [South Florida]," by S. E. Dunn. FLA ENVIRONMENTAL & URBAN ISSUES 1:7-10, October-November, 1973.

"Community noise survey of Medford, Massachusetts," by J. E. Wesler. J ACOUST SOC AM 54:985-995, October, 1973.

"Community response to elimination of nighttime aircraft noise," by H. G. Smith. J ACOUST SOC AM 55Suppl:68, 1974.

"Comparative characteristics of the action of moderate levels of constant and interrupted noise on certain body functions," by L. A. Oleshkevich GIG SANIT 38:95-97, August, 1973.

"Comparison of inside and outside noise measurements in various urban environments," by D. E. Bishop. J ACOUST SOC AM 55(2):465, 1974.

"Complex effects of different factors (noise, tranquilizer, difficult working condition, test time) on pursuit tracking performance and beat-to-beat heart rate behavior," by H. Strasser, et al. INT ARCH ARBEITSMED 31:81-103, May 23, 1973.

"Comprehensive clinical and psychological studies of patients exposed to chronic acoustic trauma," by S. Klonowski, et al. POL TYG LEK 29:313-315, February 25, 1974.

"Compressor sound control," by G. M. Diehl. TAPPI 57:75-77, April, 1974.

"Compressors beat new noise-law levels." ELEC WORLD 180:85, July 15, 1973.

"Computer program sequence for collection, reduction, analysis, and summary of auditory evoked potential data," by M. I. Mendel, et al. COMPUT BIOMED RES 6:578-587, December, 1973.

"Computerized classification of the results of screening audiometry in groups of persons exposed to noise," by K. Klockhoff, et al. AUDIOOGY 13(4):326-334, 1974.

"Conditioning treatment of enuresis: auditory intensity," by G. C. Young, et al. BEHAV RES THER 11:411-416, November, 1973.

"Conference on Vehicle noise and the designer, Hatfield, England." ENGINEER 238:27, May 2, 1974.

"Consequences of peripheral frequency selectivity for nonsimultaneous masking," by H. Duifhuis. J ACOUST SOV AM 54:1471-1488, December, 1973.

"Considerations on the tolerance of workers for ear protective devices," by J. P. Pepersack. ARCH BELG MED SOC 31:179-183, March, 1973.

"Constitutional law: aircraft noise control preempted by federal government." WASHBURN L J 13:118-123, Winter, 1974.

"Construction noise; its origin and effects," by A. S. Hersh. AM SOC C E PROC 100:433-448, September, 1974.

"Construction, transportation tabbed as noise target areas." IND W 182-18-19, July 1, 1974.

"Consultants says duct noise regeneration bothers owners, needs A/C industry study," by M. Kodaras. AIR COND HEAT & REFRIG N 130:13, October 1, 1973.

"Consumer product noise reduction," by R. S. Musa. J ENVIRONMENTAL SCI 17:9-11, July, 1974.

"Contractors: protection clause mades owner pay to quell noise complaints." AIR COND HEAT & REFRIG N 131:29, Februrary 25, 1974.

"Contrast medium studies of the internal acoustic meatus (cisternomeatography)—a system for early diagnosis of acoustic nerve tumors," by

I. Fleszar. POL PRZEGL RADIOL 38:9-17, 1974.

"A contribution to the physiology of the perilymph. 3. On the origin of noise-induced hearing loss,' by E. A. Schnieder. ANN OTOL RHINOL LARYNGOL 83:406-412, May-June, 1974.

"Control of environmental noise," by P. Jensen. J AIR POLLUT CONTROL ASSOC 23:1028-1034, December, 1973.

"Control valve muffles high-noise operations." CHEM ENG 80:56, August 6, 1973.

"Control-valve noise yields to research," by R. Nugent. POWER 117:69-71, July, 1973.

"Controlled vibration feeds dry mixes, eliminates noise problem [modern maid food products]." FOOD PROCESSING 35:93, May, 1974.

"Controlling equipment noise," by W. Murtland. RUBBER WORLD 168:65-66, July, 1973.

"Controlling in-plant noise," by K. M. Hankel. AUTOMATION 21:86-90, April, 1974.

"Controlling industrial noise; acoustic materials and enclosures," by C. H. Wick. MANUF ENG & MGT 70:30-33, June, 1973.

"Controlling industrial noise; administrative controls and hearing protection," by C. H. Wick. MANUF ENG & MGT 71:32-35, July, 1973.

"Controlling noise at work [Great Britain]," by G. R. C. Atherley, et al. LABOR RESEARCH 63:150-154, July, 1974.

"Controlling noisy washer-dryer systems." PLANT ENG 28:48, February 7, 1974.

"Controlling technically produced noise to reduce psychological stress." IMPACT OF SCIENCE ON SOCIETY 23,3:237-248, July/September, 1973.

"Convulsive seizures in autostimulation during the period of sensitivity to audiogenic seizure in DBA-2 mice," by P. Cazala, et al. C R

ACAD SCI [D] 278:2811-2814, May, 1974.

"Correlation between sound pressure and intra-thoracic pressure at onset of phonation," by F. Klingholz, et al. FOLIA PHONIATR 24:381-386, 1972.

"Cortical lesions: flavor illness and noise-shock conditioning," by W. G. Hankins, et al. BEHAV BIOL 10:173-181, February, 1974.

"A cost-effective method of evaluating aircraft noise-abatement options [San Antonio international airport]," by S. R. Goldberg. TEX BUS R 47:284-287, December, 1973.

"Cost of noise reduction in intercity commercial helicopters," by H. B. Faulkner. J AIRCRAFT 11:89-95, February, 1974.

"Crash resistance and noise improvements in business aircraft." AUTOMOTIVE ENG 82:57-59, April, 1974.

"Criteria for a recommended standard—occupational exposure to noise. I. Recommendations for a noise standard," by H. M. Utidjian. J OCCUP MED 16:33-37, January, 1974.

"Curbing noise with partial enclosures," by W. G. Phillips, et al. MACHINE DESIGN 46:107-110, April 4, 1974.

"Damage due to noise," by A. Delmas. BULL ACAD NATO MED 157(4): 272-276, 1973.

"Danger: noise at work!" by A. S. Freese. POP MECH 142:140-145 plus. November, 1974.

"Data on the hygienic evaluation of city noise," by S. A. Soldatkina, et al. GIG SANIT 38:16-20, March, 1973.

"Deafening effects of impulse noise on the rhesus monkey," ACTA OTOLARYNGOL [Suppl] 1-44, 1973.

"Decay of temporary threshold shift in noise: monaural chinchillas," by J. H. Mills, et al. J SPEECH & HEARING RES 16:267-270, June, 1973.

"Delayed learning of rats exposed to noise," by T. Mitsuya. JAP J HYG 28:324-339, August, 1973.

"Designing safety into underground mining equipment; noise abatement," by C. Holvenstot. MIN CONG J 59:39-43, September, 1973.

"Designing small gas turbine engines for low noise and clean exhaust," by H. C. Eatock, et al. J AIRCRAFT 11:616-623, October, 1974.

"Determination of in-use attenuation value for selected ear plugs," by C. E. Scott III, et al. J ACOUST SOC AM 54(1):328, 1973.

"Development of some mechanisms useful in sound localization," by S. D. Erulkar. FED PROC 33:1928-1932, August, 1974.

"Development of sonic inlets for turbofan engines," by F. Klujber. J AIRCRAFT 10:579-586, October, 1973.

"Device for the integral hygienic evaluation of noises," by P. N. Chumak, et al. GIG TR PROF ZABOL 17:48-50. July, 1973.

"Diagnosis of non-tumorous lesion of the cerebello-pontile region: with special reference to the differential diagnosis of acoustic nerve tumor," by Y. Yoshimoto. JAP J CLIN MED 31:3251-3260, November, 1973.

"Diagnosis of post-trauma vertigo," by A. Montandon. J FR OTORHINOLAYRNGOL 22:647-649, September, 1973.

"The diagnostic contribution of EEG in the assessment of noise induced hearing losses," by H. G. Dieroff, et al. Z LARYNGOL RHINOL OTOL 52:908-914, December, 1973.

"Diagnostic significance of acoustic measurement of the eyes in unilateral exophthalmos," by G. V. Kruzhkova. VESTN OFTALMOL 2:88-89, March-April, 1974.

"Diagnostic value of Bekesy comfortable loudness tracings," by J. Jerger, et al. ARCH OTOLARYNGOL 99:351-360, May, 1974.

"Dichotic competition of simultaneous tone bursts of different frequency. I. Dissociation of pitch from lateralization and loudness," by R. Efron, et al. NEUROPSYCHOLOGIA 12:249-256, March, 1974.

"Differential diagnosis of cerebello-pontine tumours," by J. Helms. LARYNGOL RHINOL OTOL 53:194-199, March, 1974.

"Differential phylogenetic development of the acoustic nuclei among chiroptera," by G. Baron. BRAIN BEHAV EVOL 9:7-40, 1974.

"Diffraction of a plane wave by a half plane in a subsonic and supersonic medium," by S. M. Candel. ACOUSTICAL SOC AM J 54:1008-1016, October, 1973.

"Directory of graduate education in acoustics," by W. M. Wright, et al. J ACOUST SOC AM 55:1105-1115, May, 1974.

"Discharge patterns of single fibers in the pigeon auditory nerve," by M. B. Sachs, et al. BRAIN RES 70:431-447, April 26, 1974.

"A 'distraction effect' of noise bursts," by S. Fisher. PERCEPTION 1:223-236, 1972.

" Distribution of the changes in the receptor auditory cells along the basilar membrane of the cochlea under the influence of narrow-band (octave) noise of different frequency characteristics," by V. F. Anichin, et al. ZH USHN NOS GORL BOLEZN 33:15-20, 1973.

"Divide and conquer your noise problems," by W. M. Ihde. FOUNDRY 101:61-62, August, 1973.

"Do not disturb [O'Hare international tower hotel]." ENGIN N 190:12, May 24, 1973.

"Does tonotopicity subserve the perceived elevation of a sound?" by R. A. Butler. FED PROC 33:1920-1923, August, 1974.

"Dolby B-type noise reduction system," by R. Berkovitz, et al. AUDIO 57:15-16, September; 33-36, October, 1973.

"Dorsal cochlear nucleus of the chinchilla: excitation by contralateral sound," by T. E. Mast. BRAIN RES 62:61-70, November 9, 1973.

"Downing the plant's din." CHEM ENG 80:30 plus, December 24, 1973.

"Ear muffs made for all-day wear." PURCHASING 77:71, December 3, 1974.

"Early averaged electroencephalic responses to clicks in neonates," by C. C. McRandle, et al. ANN OTOL RHINOL LARYNGOL 83(5):695-702, September-October, 1974.

"Early diagnosis and management of acoustic neuromas," by J. M. Tew, Jr., et al. OHIO STATE MED J 70:365-367, June, 1974.

"Early diagnosis of neurinoma of the acoustic nerve (typical and atypical forms)," by J. M. Sterkers. PROBL ACTUELS OTORHINOLARYNGOL 29-41, 1971.

"Echographic diagnosis of the ciliary body detachment," by B. N. Alekseev. VESTN OFTALMOL 4:20-27, 1973.

"Education by example?" by R. M. Merriman. OCCUP HEALTH 26:182-183, May, 1974.

"The effect of acute noise exposure on the excretion of corticosteroids, adrenalin and noredrenalin in man," by A. Slob, et al. INT ARCH ARBEITSMED 31:225-235, July 10, 1973.

"Effect of ambient illumination on noise level of groups," by M. Sanders, et al. J APP PSYCHOL 59:527-528, August, 1974.

"Effect of contralateral broad-band noise on frequency discrimination," by J. M. Labiak, et al. ACTA OTOLARYNGOL 77:29-36, January-February, 1974.

"Effect of D-amphetamine sulfate on susceptibility to audiogenic seizures in DBA-2J mice," by J. M. Graham, Jr., et al. BEHAV BIOL 10:183-190, February, 1974.

"Effect of different parameters of industrial noise on the auditory analyzer and the central nervous system of adolescent workers," by E. A. Gel'tishcheva. GIG TR PROF ZABOL 17:5-9, July, 1973.

"Effect of ejector spacing an ejector-jet noise characteristics," by D. Tirumalese. ACOUSTICAL SOC AM J 56:911-916, September, 1974.

"Effect of environment on the health of the mining personnel in copper ore mines of the Legnica-Glogow Copper Region," by I. Juzwisk, et al. POL TYG LEK 29:811-814, May 13, 1974.

"Effect of impulse noise on workers' hearing," by L. I. Maksimova, et al. GIG SANIT 38:33-36, September, 1973.

"Effect of industrial noise and ototoxic antibiotics on cochlear function," by E. Krochmalska. ACTA OTOLARYNGOL 77:44-50, January-February, 1974.

"Effect of industrial noise on the hearing organ following conservative surgery of the middle ear," by W. Sulkowski, et al. OTOLARYNGOL POL 27:617-624, 1973.

"Effect of infrasonics on the body," by E. N. Malyshev, et al. GIG SANIT 39:27-30, March, 1974.

"Effect of intensive noise on the microcirculation in the brain of experimental animals," by S. V. Alekseev, et al. GIG TR PROF ZABOL 16:24-26, July, 1972.

"Effect of an interstate highway on urban area noise levels," by J. E. Heer, Jr., et al. PUB WORKS 105:54-58, January, 1974.

"Effect of lithium carbonate and alpha-methyl-p-tyrosine on audiogenic seizure intensity," by P. C. Jobe, et al. J PHARM PHARMACOL 25: 830-831, October, 1973.

"Effect of loud music on hearing." (editorial). W VA MED J 70:165-166, July, 1974.

"Effect of noise and ototoxic substances on previously damaged ears," by M. Quante. ARCH KLIN EXP OHREN NASEN KEHIKOPFHEILKD 205:266-269, December 17, 1973.

"Effect of noise exposure during primary flight training on the conventional and high frequency hearing of naval aviation officer candidates," by R. M. Robertson, et al. J ACOUST SOC AM 55:Suppl:41, 1974.

"Effect of noise on the ear following tympanoplasty in miners employed underground," by S. Stawinski. OTOLARYNGOL POL 27:751-755, 1973.

"Effect of noise on the general immunological reactivity of the body," by M. L. Khaimovich. GIG SANIT 38:96-98, February, 1973.

"The effect of noise on human sensations," by L. A. Oleshkevich. VRACH DELO 3:127-130, 1973.

"Effect of noise on information processing processes and on simple motor reaction," by B. Stefenov, et al. GIG SANIT 37:84-86, September, 1972.

"Effect of noise on intellectual performance," by N. D. Weinstein. J APP PSYCHOL 59:548-554, October, 1974.

"Effect of noise on man," by A. Mann. HAREFUAH 83:387-388, November 1, 1972.

"Effect of noise on the Stroop Test," by L. R. Hartley, et al. J EXP PSYCHOL 102:62-66, January, 1974.

"The effect of noise on visual fields," by J. E. Letourneau, et al. EYE EAR NOSE THROAT MON 53:49-51, February, 1974.

"Effect of priming and testing for audiogenic seizures in BALB-c mice as a function of stimulus intensity," by C. S. Chen, et al. EXPERIENTIA 30:153, February 15, 1974.

"Effect of sound on creative performance," by B. Kaltsounis. PSYCHOL REP 33:737-738, December, 1973.

"The effect of vibration and noise on development of inflammatory reaction in rats," by J. Billewicz-Stankiewicz, et al. ACTA PHYSIOL POL 25:235-240, May, 1974.

"Effect of white noise on the reaction time of mentally retarded subjects," by C. M. Miezejeski. AM J MEN DEFICIENCY 79:39-43, July, 1974.

"The effect of white noise on the somatosensory evoked response in sleeping newborn infants," by P. H. Wolff, et al. ELECTROENCEPHALOGR CLIN NEUROPHYSIOL 37:269-274, September, 1974.

"Effectiveness of different ear protectors in protecting the employee from over exposure in industrial environments," by J. E. Stephenson, et al. J ACOUST SOC AM 54(1):301, 1973.

"Effects of acoustical stimulation on equilibrium," by S. L. Vanderhei.

J ACOUST SOC AM 55:Suppl:41, 1974.

"Effects of airplane noise on health: an examination of three hypotheses," by D. B. Graevan. J HEALTH & SOC BEHAV 15:336-343, December, 1974.

"Effects of frequency modulation on auditory averaged evoked response," by M. L. Lenhardt. AUDIOLOGY 10:18-22, January-February, 1971.

"Effects of industrial noise upon the hearing organs following radical surgery of the middle ear," by J. Laciak, et al. OTOLARYNGOL POL 27:485-491, 1973.

"The effects of intelligence quotient and extraneous stimulation on incidental learning," by R. Forehand, et al. J MENT DEFIC RES 17:24-27, March, 1973.

"Effects of intense auditory stimulation: hearing losses and inner ear changes in the chinchilla," by I. M. Hunter-Duvar, et al. J ACOUST SOC AM 55:795-801, April, 1974.

"Effects of intense auditory stimulation: hearing losses and inner ear changes in the squirrel monkey. II.," by I. M. Hunter-Duvar, et al. J ACOUST SOC AM 54:1179-1183, November, 1973.

"Effects of intermittent, moderate intensity noise stress on human performance," by G. C. Theologus, et al. J APP PSYCHOL 59:539-547, October, 1974.

"The effects of intermittent noise on human serial docoding performance and physiological response," by D. W. Conrad. ERGONOMICS 16:739-747, November, 1973.

"Effects of low-frequency vibration and noise on conditioned avoidance reaction in rats," by J. Billewicz-Stankiewicz, et al. ACTA PHYSIOL POL 25(4):307-312, July-August, 1974.

"Effects of low-pass filtering on the rate of learning and retrieval from memory of speech-like stimuli," by R. Novak, et al. J SPEECH HEAR RES 17:279-285, June, 1974.

"Effects of masking-spectrum slope and interaural phase on detection of tones," by M. Sonn. PERCEPT MOT SKILLS 38:776-784, June, 1974.

"The effects of noise level and elevated ambient temperatures upon selected reproductive traits in female Swiss-Webster mice," by H. B. Zekem, et al. LAB ANIM SCI 24:469-475, June, 1974.

"Effects of noise on people," by J. D. Miller. ACOUSTICAL SOC AM J 56:729-764, September, 1974.

"The effects of noise on the performance of simultaneous interpreters: accurarcy of performance," by D. Gerver. ACTA PSYCHOL 38(3): 159-167, June, 1974.

"Effects of noise pollution on animal behavior," by W. E. Brewer. CLIN TOXICOL 7:179-189, April, 1974.

"Effects of a traffic noise background on judgments of aircraft noise," by C. A. Powell. J ACOUST SOC AM 55:Suppl:68, 1974.

"Effects of varying levels of interruption of temporary threshold shift," by M. E. Schmidek. J ACOUST SOC AM 55:Suppl:40, 1974.

"Effects of white noise on the frequency of stuttering," by S. F. Garber, et al. J SPEECH & HEARING RES 17:73-79, March, 1974.

"The effects of white noise on PB scores of normal and hearing-impaired listeners," by R. W. Keith, et al. AUDIOLOGY 11:177-186, May-August, 1972.

"The effects of temporary hearing loss with combined impact and steady state noise," by K. Yamamura, et al. JAP J HYG 28(6):517-521, February, 1974.

"Efforts continue on reducing noise, ink, dust in pressroom," by G. B. Healey. ED & PUB 107:10, January 26, 1974.

"Elastomer spray cuts plant noise." PURCHASING 76:57, January 8, 1974.

"Election kills noise bill (Environment Protection Bill)." MEL MAKER 49:5, February 16, 1974.

"Electrocochleography," by F. B. Simmons. ANN OTOL RHINOL LAR-

YNGOL 83:312-313, May-June, 1974.

"Electrocochleography in clinical-audiological diagnosis," by H. Sohmer, et al. ARCH OTORHINOLARYNGOL 206:91-102, March 25, 1974.

"Electrographic correlates of lateral asymmetry in the processing of verbal and nonverbal auditory stimuli," by H. Neville. J PSYCHOLINGUIST RES 3:151-163, April, 1974.

"Electronic reverberation equipment in the Stockholm concert hall," by S. Dahlstedt. AUDIO ENG SOC J 22:627-631, October, 1974.

"Engine-over-the-wing noise research," by M. Reshotko, et al. J AIRCRAFT 11:95-96, April, 1974.

"Environment resource packets get wide use." CHEMICAL AND ENGINEERING NEWS 52,4:25-26, January, 1974.

"Environmental law—aircraft noise control—use of local police powers to impose curfews on air flights is pre-empted by the federal aviation act of 1958 as amended by the noise control act of 1972." RUTGERS CAMDEN L J 5:566-584, Spring, 1974.

"Environmental law—aircraft noise regulation—federal pre-emption." NY L F 20:165-176, Summer, 1974.

"Environmental law—federal regulation of aircraft noise under federal aviation act precludes local police power noise restrictions." BC IND & COM L R 15:848-862, April, 1974.

"Environmental law—federal regulation under federal aviation act and noise control act preempts the field of airport and aircraft noise control rendering local airport curfews invalid." KAN L REV 22:319-336, Winter, 1974.

"Environmental law: the noise control act of 1972." OKLA L REV 27:55-62, Winter, 1974.

"Environmental noise," by F. Merluzzi. MED LAV 64:115-120, March–April, 1973.

"Environmental noise classifier and its use," by J. Donovan. AUDIO ENG

SOC J 22:528-532, September, 1974.

"Environmental noise impact of natural draft hyperbolic cooling towers," by J. P. Carlson, et al. J ACOUST SOC AM 55:Suppl:36, 1974.

"Environmental noise level as a factor in the treatment of hospitalized schizophrenics," by M. F. Ozerengin. DIV NERV SYST 35(5):241-245, 1974.

"Environmental noise management," by D. P. Loucks, et al. AM SOC C E PROC 99:813-829, December, 1973.

"Environmental noise; status of the agency programs." ASTM STAND N 2:32 plus, May, 1974.

"EPA proposes rail noise standards." AIR POLLUTION CONTROL ASSN J 24:881, September, 1974.

"EPA readies noise recommendations," by R. K. Ellingsworth. AVIATION W 100:32-33, March 25, 1974.

"EPA report lists targets for noise standards." MACHINE DESIGN 46:6, July 25, 1974.

"Equal aversion levels for pure tones and one third-octave bands of noise," by J. A. Molino. J ACOUST SOC AM 55:1285-1289, June, 1974.

"Equating individual differences for auditory input," by D. McGuinness. PSYCHOPHYSIOLOGY 11:113-120, March, 1974.

"Establishment of noise level standards in administrative and public buildings," by V. A. Tokarev, et al. GIG TR PROF ZABOL 0(8):13-16, August, 1974.

"Etiological factors in hearing loss in tangential missile wounds of the head," by A. Adeloye, et al. LARYNGOSCOPE 84:126-131, January, 1974.

"Europe seeks common solutions to problems of emissions and noise." AUTOMOTIVE ENG 81:25-31, March, 1973.

"Evaluation of the decrease in auditory function in railroad workers according to tonal audiographic data," by I. E. Zaslavskii. GIG TR PROF ZABOL 16:44-47, July, 1972.

"Evoked response thresholds for long and short duration tones," by C. T. Grimes, et al. AUDIOLOGY 10:358-364, September-December, 1971.

"Expensive sound of silence." BUS W 28, July 20, 1974.

"Experiment on electronic noise in the freshman laboratory," by D. L. Livesey, et al. AM J PHYS 41:1363-1367, December, 1973.

"Exposure of the internal auditory meatus by the House-Fisch-Portmann method with transsection of part of the 8th nerve," by B. Latkowski, et al. OTOLARYNGOL POL 27:569-575, 1973.

"The exposure of truck drivers to noise," by B. H. Sharp. J ACOUST SOC AM 55(2):484, 1974.

"Extraversion and auditory sensitivity to high and low frequency," by R. M. Stelmack, et al. PERCEPT MOT SKILLS 38:875-879, June, 1974.

"FAA gets contradictory noise guidance," by W. A. Shumann. AVIATION W 101:40-42, August 12, 1974.

"FAA noise proposal draws big response from citizen groups." AVIATION W 101:26, July 22, 1974.

"FAA proposes to quiet jets." ASTRONAUTICS & AERONAUTICS 12:13, May, 1974.

"FAA pushing nacelle noise retrofits," by W. A. Shumann. AVIATION W 100:29-30, May 27, 1974.

"Fallacies of silence," by H. Carruth. HUDSON R 26:462-470, Autumn, 1973.

"Federal pre-emption and airport noise control." URBAN L ANN 8:229-239, 1974.

"Federal pre-emption—aviation noise control—the Federal aviation administration, monitored by the Environmental protection agency, has full control over aviation noise, pre-empting state and local control, including a municipal ordinance which imposed a curfew on certain jet take-offs during certain night-time hours." J AIR L 40:341-349, Spring, 1974.

"Federal preemption in airport noise abatement regulation: allocation of federal and state power." MAINE L REV 26:321-344, 1974.

"Field investigation by hypnosis of sound loci importance in human behavior," by M. H. Erickson. AM J CLIN HYPN 16:92-109, October, 1973.

"First plug your ears." ECONOMIST 248:49-50, September 1, 1973.

"First result of the noise control act; how noise affects people," by R. A. Jacobson. MACHINE DESIGN 45:132-136, October 18, 1973.

"For goodness sake—let your patients sleep!" by D. A. Grant, et al. NURSING 4:54-57, November, 1974.

"For noise control, absorption method called least effective." AIR COND HEAT & REFRIG N 133:20, December 9, 1974.

"Four leads slash capacitor noise." MACHINE DESIGN 46:47, August 8, 1974.

"Four ways materials combat noise pollution," by K. H. Miska. MATERIALS ENG 79:20-23, June, 1974.

"Fragmented noise control sought; EPA to press for noise funding scheme," by C. E. Schneider. AVIATION W 99:19-20, July 9; 32-33, July 16, 1973.

"F28; development of the MK 5000/6000; further development of the MK 1000." AIRCRAFT ENG 45:14-19, October, 1973.

"Functional and histological findings in acoustic tumor," by M. Igarashi, et al. ARCH OTOLARYNGOL 99:379-384, May, 1974.

"Functional state of the auditory analyzer under conditions of prolonged clinostatic hypokinesia," by Z. I. Matsnev. VOEN MED ZH 7:62-65, July, 1973.

"Functional state of the hearing analyzer in the concurrent action of noise, vibration, physical work and high temperature," by M. V. Ratner, et al. GIG TR PROF ZABOL 16:47-49, July, 1972.

"Functional tests in lesions of the trigeminal n (V), acoustic n (V3) and the nervus intermedius in the region of the cerebello-pontine angle after Dandy's operation," by U. Koch, et al. Z LARYNGOL RHINOL OTOL 52:729-736, October, 1973.

"Further data, contributed by electrocochleography, on the funtion of normal and pathological peripheral receptors," by J. M. Aran. ACTA OTORHINOLARYNGOL BELG 26:671–683, 1972.

"Further studies on industrial sudden deafness," by F. Suga. J OTOLARYNGOL JAP 76:1373-1379, November, 1973.

"Future fans: big and quiet." OIL & GAS J 72:71, June 10, 1974.

"Gammagraphic diagnosis in tumors of the 8th pair," by J. M. Sampere, et al. ARCH NEUROBIOL 37:45-60, January-February, 1974.

"Gear noise source indentification and reduction," by R. F. MacWhorter. AM IND HYG ASSOC J 35(9):581-585, September, 1974.

"Gear pump noise," by R. C. Michel. FUELOIL & OIL HEAT 33:46 plus, July, 1974.

"Geared fan engine systems; their advantages and potential reliability," by T. A. Lyon, et al. J AIRCRAFT 10:361-365, June, 1973.

"A gene controlling bell- and photically-induced ovulation in mice," by B. E. Eleftheriou, et al. J REPROD FERTIL 38:41-47, May, 1974.

"Generality of interference by tonal stimuli in recognition memory for pitch," by D. Deutsch. Q J EXP PHYSIOL 26:229-234, May, 1974.

"Generalization of stimulus control in a summer camp," by G. E. Taylor, Jr., et al. PSYCHOL REPT 34:419-423, April, 1974.

"Get the lead in, get noise out [sound–insulated enclosures]." FACTORY 6:11, November, 1973.

"Getting noise immunity in industrial controls," by H. M. Schlicke, et al. IEEE SPECTRUM 10:30-35, June, 1973.

"GM's noise program: broad, wide and deep," by C. A. Gottesman. AUTOMOTIVE IND 151:54 plus, November 1, 1974.

"Graphic method of determining the distance from a noise source to the area the level is standardized," by A. L. Vasil'eva, et al. GIG SANIT 38:84-86, March, 1973.

"Guide to control valve noise: with chart," by J. A. Dillons. INSTR & CONTROL SYSTEMS 47:95-104, September, 1974.

"Guidelines for noise control." AM DYESTUFF REP 62:59-63, July; August, 1973.

"Guidelines for textile industry noise control," by J. R. Bailey, et al. J ENG IND 96:241-246, February, 1974.

"The hazardous noise exposure to which airline passengers are subjected," by S. R. Lane. J ACOUST SOC AM 55(2):465, 1974.

"Health evaluation of the noise of automotive agricultural machines," by V. V. Vlasenko. GIG TR PROF ZABOL 16:52-54, July, 1972.

"Health hazard: sound pollution." MUS J 31:27 plus, December, 1973.

"A hearing conservation programme." HEALTH PEOPLE 8:8-9, March, 1974.

"Hearing damage from music: United Kingdom experience," by A. Burd. AUDIO ENG SOC J 22:524-527, September, 1974.

"The hearing disability of the noise-damaged and the industrial injury insurance," by I. Klockhoff, et al. LAKARTIDNINGEN 71:819-822, February 27, 1974.

"Hearing loss among Baffin Zone Eskimos—a preliminary report," by J. D. Baxter, et al. CAN J OTOLARYNGOL 1:337-343, 1972.

"Hearing loss due to minibikes," by R. P. Oppenheimer, et al. AM FAM PHYSICIAN 8:125, October, 1973.

"Hearing loss in adults: relation to age, sex, exposure to loud noise, and cigarette smoking," by A. B. Siegelaub, et al. ARCH ENVIRON HEALTH 29:107-109, August, 1974.

"Hearing loss: ways to avoid it, or live with it. An interview," by R. E. Jordan. US NEWS 76:48-50, January 21, 1974.

"Helicopter noise experiments in an urban environment," by W. A. Kinney, et al. ACOUSTICAL SOC AM J 56:332-337, August, 1974.

"Helplessness, stress level, and the coronary-prone behavior pattern," by D. S. Krantz, et al. J EXP SOC PSYCHOL 10:284-300, May, 1974.

"High frequency attenuation characteristics of ear protectors," by F. H. Bess, et al. J ACOUST SOC AM 54(1):328, 1973.

"High frequency attenuation characteristics of ear protectors," by T. H. Townsend, et al. J OCCUP MED 15:888-891, November, 1973.

"High-intensity ultrasonic sound: a better rat," by B. J. Morley, et al. PSYCHOL REPT 35:152-154, August, 1974.

"High noise limits will hit British truck men." ENGINEER 237:1, August 16, 1973.

"Highway noise and acoustical buffer zones," by A. Zulfacar, et al. AM SOC C E PROC 100:389-401, May, 1974.

"Histochemical activity of succinate dehydrogenase in guinea pig cochlea after impulse stimulation," by H. Guttmacher, et al. ACTA OTOLARYNGOL 76:323-327, November, 1973.

"Hitselberger' sign-its significance in the diagnosis of acoustic neurinoma," by H. Veidauer, et al. ARCH KLIN EXP OHREN NASEN KEHIKOPF-HEILKD 205:126-130, December 17, 1973.

"Holding down the decibel count." PLASTICS ENG 30:23, April, 1974.

"Home builder group rents out noise measuring instruments," by K. MacDonald. AIR COND HEAT & REFRIG N 130:37, October 29, 1973.

"Homolateral and contralateral masking of tinnitus by noise-bands and by pure tones," by H. Feldmann. AUDIOLOGY 10:138-144, May-June, 1971.

"Hospital noise." (letter). N ENGL J MED 290:522-523, February 28, 1974.

"Hospital noise." NURS DIGEST 2:61, May, 1974.

"Hospital noise may cause patient problems." J ENVIRON HEALTH 36: 354, January-February, 1974.

"Hospital noises can turn frightened patients into terrified patients," by W. A. Nolen. AMER MED NEWS 17:13, January 7, 1974.

"Hospital tranquility starts with mechanical systems: Nash General Hospital, Rocky Mount, N.C. HEATING PIPING AIR CONDITIONING 45:24, February, 1973.

"House unit urges FAA to delay rule on quiet nacelle retrofit." AVIATION W 100:24, June 24, 1974.

"How Day & Night-Payne reduced sound, increased efficiencies." AIR COND HEAT & REFRIG N 131:27, February 18, 1974.

"How R. J. Reynolds protects workers' hearing." MGT R 62:64-66, July, 1973.

"How sound an environment?" by M. Cooper. DAILY TELEGRAPH 11, May 4, 1974.

"How to rate noise sources." MACHINE DESIGN 46:152, May 16, 1974.

"How wind noise affects human hearing." MACHINE DESIGN 45:6, November 1, 1973.

"H-P noise monitor fails, repay $178K to LA airport," by J. Fraser. ELECTRONIC N 19:1 plus, March 25, 1974.

"Human perception of apparent direction and movement of aircraft noise," by W. J. Gunn, et al. J ACOUST SOC AM 55:Suppl:68, 1974.

"Human temporary threshold shift from 16-hour noise, exposures," by

W. Melnick. ARCH OTOLARYNGOL 100:180-189, September, 1974.

"Hygienic characteristics of noise and an analysis of the disease incidence among workers engaged in the metalworking industry," by E. P. Orlovskaia, et al. VRACH DELO 7:121-125, July, 1973.

"Hygienic working conditions and state of health of operators of track maintenance and repair machines," by E. I. Gol'Dman, et al. GIG TR PROF ZABOL 17(10):45-46, 1973.

"Hypertensive effects of prolonged auditory, visual, and motion stimulation," by H. H. Smookler, et al. FED PROC 32:2105-2110, November, 1973.

"Identification of renal calculi by a new sonar blunt curette," by H. Tammen. UROL INT 28:158-160, 1973.

"If you can't beat noise, baffle it." ENGINEER 237-23, September 13, 1973.

"Implementation of a hearing preservation program," by M. T. Summar. AM PAPER IND 55:16-18 plus, November, 1973.

"Impulse noise trauma. A study of histological susceptibility," by R. P. Hamernik, et al. ARCH OTOLARYNGOL 99:118-121, February, 1974.

"Increased adult auditory responsiveness resulting from juvenile acoustic experience," by K. R. Henry. FED PROC 32:2098-2100, November, 1973.

"Increased noise as an element of compensation in condemnation proceedings," by W. R. Theiss. APPRAISAL J 42:134-138, January, 1974.

"Individual hearing protection-survey of Berlin's noisy factories," by P. Moch. ZENTRALBL ARBEITSMED 23:33-38, February, 1973.

"Industrial medicine. Mass screening of workers exposed to noise," (proceedings), by K. Humperdinck, et al. HEFTE UNFALLHEILKD 114: 193-197, 1973.

"Industrial noise and hearing loss," by C. O. Istre, Jr., et al. J LA STATE

MED SOC 126:5-7, January, 1974.

"Industrial noise and vibration in sewing industry enterprises and an evaluation of measures to decrease them," by V. F. Rudenko, et al. GIG TR PROF ZABOL 17:36-38, July, 1973.

"Infrasound." (editorial). LANCET 2:1368-1369, December 15, 1973.

"Infrasound-occurrence and effects," by S. Handel, et al. LAKARTIDNINGEN 71:1635-1639, April 17, 1974.

"Inhibition of fetal osteogenesis by maternal noise stress," by W. F. Geber. FED PROC 32:2101-2104, November, 1973.

"In-place machinery noise measurements," by C. E. Ebbing, et al. ASHRAE J 15:48-54, June, 1973.

"Instrumentation for measuring and analyzing noise," by T. T. Weissenburger. PLANT ENG 27:80-84, November 1, 1973.

"Insulating against jet noise." CHEMISTRY 47:21, February, 1974.

"Intact vestibular and cochlear function in acoustic neuroma," by W. S. Gunasekera, et al. CEYLON MED J 18:113-115, June, 1973.

"Interaction of continuous and impulse noise: audiometric and histological effects," by R. P. Hamernik, et al. J ACOUST SOC AM 55:117-121, January, 1974.

"Interactions and range effects in experiments on pairs of stresses: mild heat and low-frequency noise," by E. C. Poulton, et al. J EXP PSYCHOL 102:621-628, April, 1974.

"Interaural alternation and speech intelligibility," by C. Speaks, et al. J ACOUST SOC AM 56(2):640-644, August, 1974.

"Interdisciplinary plant-noise control," by A. Thumann. CHEM ENG 81: 120 plus, August 19, 1974.

"Intermittent noise exposure and associated damage risk to hearing of chain saw operators," by M. Schmidek, et al. AM IND HYG ASSOC J 35: 152-158, March, 1974.

"International standardization for noise." ASTM STAND N 2:50, May, 1974.

"Intersensory facilitation, errors and corrections in a discontinuous tracking task," by A. Semjen. ANNEE PSYCHOL 73:403-417, 1973.

"Intracellular electric responses to sound in a vertebrate cochlea," by M. J. Mulroy, et al. NATURE 249:482-485, May 31, 1974.

"Intracellular septate desmonsome-like structures in a human acoustic Schwannoma in vitro," by F. K. Conley, et al. J NEUROCYTOL 2: 457-464, December, 1973.

"Inverse condemnation and nuisance: alternative remedies for airport noise damage." SYRACUSE L REV 24:793-809, 1973.

"Is frequency information extracted from electrical stimulation of the auditory system?" by F. W. Mis, et al. EXP NEUROL 43:227-241, April, 1974.

"Isotopic and neuroradiologic correlation in the examination of auditory nerve neurinoma," by C. Bamberger-Bozo, et al. ROENTGENBLAETTER 26:182-189, April, 1973.

"Jaundiced eye," by S. Novick. ENVIRONMENT 15:inside cover, December, 1973.

"Jet engine exhaust noise due to rough combustion and nonsteady aerodynamic sources," by E. G. Plett, et al. ACOUSTICAL SOC AM J 56: 516-522, August, 1974.

"Jet noise at schools near Los Angeles International Airport," by S. R. Lane, et al. J ACOUST SOC AM 56:127-131, July, 1974.

"Job safety-national compliance pact." FOOD PROCESSING 35:10, June, 1974.

"Judged acceptability of noise exposure during television viewing," by L. E. Langdon, et al. ACOUSTICAL SOC AM J 56:510-515, August, 1974.

"Key noise reduction decisions imminent," by W. A. Shumann. AVIATION W 100:31-33, January 7, 1974.

"Kinematic sound screen; unique solution to highway noise abatement," by J. B. Hauskins, Jr. AM SOC C E PROC 100:169-178, February, 1974.

"Laboratory note. Scalp-recorded early responses in man to frequencies in the speech range," by G. Moushegian, et al. ELECTROENCEPHALOGR CLIN NEUROPHYSIOL 35:665-667, December, 1973.

"Labyrinthectomy: indications, technique and results," by J. L. Pulec. LARYNGOSCOPE 84(9):1552-1573, September, 1974.

"Large gas handling plants in noise control," by T. Dear. CHEM ENG PROG 70:65-68, February, 1974.

"Leakage path of room noise to the phonocardiographic microphones," by N. Suzumura, et al. JAP J MED ELECTRON 11:344-349, October, 1973.

"Lesions in the septal nuclei of the rat raise mean systemic arterial pressure and prevent the development of sound-withdrawal hypertension," by J. F. Marwood, et al. J PHARM PHARMACOL 25:614-620, August, 1973.

"Lighting, climate, and acoustics in large office rooms," by F. Roedler. ZENTRALBL BAKTERIOL 225:316-328, December, 1973.

"Localization of acoustic stimulation in fishes and amphibia," by E. Schwartz. FORTSCHR ZOOL 21:121-135, 1973.

"Looking for noise?" FACTORY 7:54-55, February, 1974.

"Loss and recovery processes operative at the level of the cochlear microphonic during intermittent stimulation," by G. R. Price. J ACOUST SOC AM 56:183-189, July, 1974.

"Loudness discomfort level: selected methods and stimuli," by D. E. Morgan, et al. J ACOUST SOC AM 56(2):577-581, August, 1974.

"Loudness discomfort level under earphone and in the free field: the effects of calibration methods," by D. E. Morgan, et al. J ACOUST SOC AM 56:172-178, July, 1974.

"Loudness of brief tones in hearing-impaired ears. Temporal integration of acoustic energy at suprathreshold levels in patients with presbyacusis," by C. B. Pedersen, et al. ACTA OTOLARYNGOL 76:402-409, December, 1973.

"Loudness tracking and the staircase method in the measurement of adaptation," by T. E. Stokinger, et al. AUDIOLOGY 11:161-168, May-August, 1972.

"Low-cost approach to area-wide noise monitoring," by T. E. Siddon, et al. J ACOUST SOC AM 54:646-649, September, 1973.

"Lowering Diesel noise through hardware modifications; fleet week preview." AUTOMOTIVE ENG 81:41-47, June, 1973.

"Lowering equipment noise levels; question and answers." ELEC CONSTR & MAINT 72:142, August, 1973.

"Machine noise is reduced by fitting sliding shutters." ENGINEER 238:23, June 6, 1974.

"Machinery noise control," by R. T. Booth. OCCUP HEALTH 26:52-53, February, 1974.

"Man and the environmental noise," by E. Wende, et al. INTERNIST 14:224-229, May, 1973.

"Man-made noise in urban environments and transportation systems: models and measurements," by D. Middleton. IEEE TRANS COM 21:1232-1241, November, 1973.

"Marked acoustical signs of voice virilization in girls," by A. Pruszewicz, et al. FOLIA PHONIATR 25:331-341, 1973.

"The masking noise and its effect upon the human cortical evoked potential," by H. G. Ghueden. AUDIOLOGY 11:90-96, January-April, 1972.

"The masking of binaural beats of a pure sound with a differential sound," by R. Piazza. AUDIOLOGY 11:169-176, May-August, 1972.

"Masking produced by sinusoids of slowly changing frequency," by D. A. Ronken. J ACOUST SOC AM 54:905-915, October, 1973.

"Massport vs. community; Logan international airport expansion controversy," by D. Nelkin. SOCIETY 11:27-36 plus, May, 1974.

"Materials that build a box around noise," by B. D. Wakefield. IRON AGE 214:53-54 plus, July 15, 1974.

"Maze silences valve noise." MACHINE DESIGN 46:38, September 19, 1974.

"Measured variations in aircraft noise near Arlanda airport,"by A. R. Kajland. ACOUSTICAL SOC AM J 56:329-331, August, 1974.

"Measurement of 'instantaneous' carrier frequency of bat pulses," by P. J. Kindlmann, et al. J ACOUST SOC AM 54:1380-1382, November, 1973.

"The measurement of noise from moving vehicles," by E. L. Hixson, et al. J ACOUST SOC AM 54(1):332, 1973.

"Measurement of temporal shifting of hearing threshold as an evaluation of hearing loss risk under industrial conditions," by T. Malinowski, et al. OTOLARYNGOL POL 28:167-172, 1974.

"Measures to combat noise must allow for the public good," by P. McCallum. ENGINEER 238:36-37, April 18, 1974.

"Measuring procedure for urban noise in the center of Mexico City," by F. Groenwold, et al. J ACOUST SOC AM 55(2):465, 1974.

"Mechanical equipment noise and vibration control," by L. F. Yerger. HEATING PIPING 45:61-66, July, 1973.

"Medical treatment of deafness," by J. P. Secretan. REV MED SUISSE ROMANDE 93:975-977, December, 1973.

"Medico-legal aspects of noise," by R. Murray. OCCUP HEALTH 25:55-59, February, 1973.

"Meeting, 86th, Los Angeles, October 30-November 2: program and abstracts of papers." ACOUSTICAL SOC AM J 55:383-492, February, 1974.

"Metalworking noise is a pain in the ear," by B. D. Wakefield. IRON AGE 212:55-60, December 13, 1973.

"A method for the assessment of impact noise with respect to injury to hearing," by A. M. Martin, et al. ANN OCCUP HYG 16:19-26, April, 1973.

"Method of composing a noise map of a city," by O. S. Rastorguev, et al. GIG SANIT 37:62-65, October, 1972.

"Method of improving the noise stability of a magnetic recording system in registering biomedical information," by O. V. Balabanov, et al. VESTN AKAD MED NAUK SSSR 28:61-64, 1973.

"A methodology for assessing potential community impact resulting from noise emitted by railroad yard operations," by J. W. Swing, et al. J ACOUST SOC AM 55(2):465-466, 1974.

"Methods of acoustical analysis of speeech," by M. Wajskop. ACTA OTO-RHINOLARYNGOL BELG 26:741-756, 1972.

"Mexican laws regarding control of smoke dust and noise pollution and their effects upon the oil milling industry," by M. Castaneda. J AM OIL CHEM SOC 51(2):279A, 1974.

"Microsurgery of the internal auditory canal," by J. M. Sterkers. PROBL ACTUELS OTORHINOLARYNGOL 75-88, 1972.

"Microwave hearing: evidence for thermoacoustic auditory stimulation by pulsed microwaves," by K. R. Foster, et al. SCIENCE 185:256-258, July 19, 1974.

"Middle ear measurements," by G. T. Wolcott, et al. J MED ASSOC STATE ALA 43:496-498, February, 1974.

"Milwaukee adopts noise law for home air conditioners." AIR COND HEAT & REFRIG N 130-1 plus, September 3, 1973.

"MM11 at the 86th meeting of ASA," by S. R. Lane. J ACOUST SOC AM 55:1346-1348, June, 1974.

"Model for mechanical to neural transduction in the auditory receptor," by M. R. Schroeder, et al. J ACOUST SOC AM 55:1055-1060, May, 1974.

"Model for wave propagation in a lossy vocal tract," by M. M. Sondhi. J ACOUST SOC AM 55:1070-1075, May, 1974.

"Modification of the rat's startle reaction by an antecedent change in the acoustic environment," by C. L. Stitt, et al. J COMP PHYSIOL PSYCHOL 86:826-836, May, 1974.

"Modifications of epinephrine, norepinephrine, blood lipid fractions and the cardiovascular system produced by noise in an industrial medium," by G. A. Ortiz, et al. HORM RES 5:57-64, January, 1974.

"Monaural and binaural speech perception through hearing aids under noise and reverberation with normal and hearing-impaired listeners," by A. K. Nabelek, et al. J SPEECH & HEARING RES 17:724-739, December, 1974.

"Monitoring community noise," by M. C. Branch, et al. AM INST PLAN J 40:266-273, July, 1974.

"Monkeys agree-noise is upsetting." MED TIMES 102:71-72 plus, March, 1974.

"Monosynaptic projections from the pontine reticular formation to the 3rd nucleus in the cat," by S. M. Highstein, et al. BRAIN RES 75:340-344, July 26, 1974.

"More phenomena," by C. T. T. Comber. ELECTRONICS & POWER 19: 298, July 12, 1973.

"Morphologic changes in the nerve cells of the rabbit brain caused by industrial noise," by J. Tarmas, et al. FOLIA MORPHOL 33:5-12, 1974.

"Morphological changes in the cochlea in experimental noise trauma: phase contrast microscopy," by E. Sh. Suladze, et al. VESTN OTORINOLARYNGOL 35:23-26, May-June, 1973.

"Morphological differentiation of nerve cells in rats of two strains with different genetically determined reactions to sound," by I. Y. Raushenbahk, et al. SOV J DEV BIOL 3:133-138, March-April, 1972.

"Morphology and function of the dorsal sound producing scales in the tail of Teratoscincus scincus (Reptilla; Gekkonidae)," by U. Hiller. J MOR-

PHOL 144(1):119-130, September, 1974.

"Motorway noise and dwellings." BUILD RES ESTAB DIGEST 153:1-7, May, 1973.

"MTIRA turns up volume on tool noise research." ENGINEER 239:9, May 30, 1974.

"Muffling techniques for reducing pneumatic tool noise," by R. A. Willoughby, et al. PLANT ENG 27:109-111, September 6, 1973.

"NASA JT8D refan program nears end; noise reductions achieved with refanned jet engines," by M. L. Yaffee. AVIATION W 101:46-47, July 1974.

"Need for noise control." METALLURGIA & METAL FORMING 40:270, September, 1973.

"Neurilemmona of the ninth cranial nerve masquerading as an acoustic neuroma," by J. R. Mountjoy, et al. ARCH OTOLARYNGOL 100:65-67, July, 1974.

"Neurinoma of the acoustic nerve," by G. Djupesland. TIDSSKR NOR LAEGEFOREN 93:1751-1753, September 10, 1973.

"Neurophysiological mechanisms of sound localization," by A. Starr. FED PROC 33:1911-1914, August, 1974.

"Neurosensory hypoacusia in metallurgy workers," by R. Benavides. REV MED CHIL 101:613-620, August, 1973.

"New automatic noise-reduction system (ANRS)," by M. Yamazaki, et al. AUDIO ENG SOC J 21:445-449, July-August, 1973.

"New concepts for the open office," by D. Meisner. ADM MGT 35:22-24 plus, March, 1974.

"New interpretation of noise reduction by matching," by Y. Nezer. IEEE PROC 62:404-406, March, 1974.

"New materials reduce noise, vibration." AUTOMOTIVE ENG 81:10, March, 1973.

"New method of diagnosis of joint diseases-arthrophonography," by M. A. Iasinovskii, et al. KLIN MED 51:25-28, July, 1973.

"New multipurpose Hall; theater and broadcasting facilities," by H. Moriyama. SMPTE J 83:169-175, March, 1974.

"New noise exposure level may become OSHA standard." AUTOMATION 20:12, October, 1973.

"New offender; intermittent noise," by B. D. Wakefield. IRON AGE 213: 46-48, April 29, 1974.

"New valve-silencer reduces noise level on mill steam line [Ontario-Minnesota pulp and paper]." PULP & PA 48:122, May, 1974.

"New way to lower pressroom noise level," by J. Hennage. ASSE J 19: 44-46, May, 1974.

"New York State contruction noise survey," by D. A. Driscoll, et al. J ACOUST SOC AM 55:Suppl:37, 1974.

"Next ten years in truck technology; noise." AUTOMOTIVE ENG 81:37 plus, June, 1973.

"No place to land: the airport crisis." READ DIGEST 102:101-105, March, 1973.

"Noise." ROY TOWN PLAN INST J 59:7-16, January, 1973.

"Noise," by W. J. Gould, et al. ANN NY ACAD SCI 216:17-29, 1973.

"Noise," by L. Rosenhouse. NURSING CARE 7:26-28, November, 1974.

"Noise abatement," by P. Kelsey. AIR COND HEAT & REFRIG N 132: 1 plus, July 15, 1974.

"Noise abatement, problems and progress, symposium." DIESEL EQUIP SUPT 52:42-44, June, 1974.

"Noise abatement program," by M. H. Miller. J ACOUST SOC AM 54(1): 288, 1973.

"Noise and the highway patrolman," by W. R. Pierson, et al. J OCCUP MED 15:892-893, November, 1973.

"Noise and the navy." MED TIMES 102:83, March, 1974.

"Noise and pile driving," by D. J. Hagerty, et al. ROADS & STS 117:70-71, August, 1974.

"Noise and the truck drivers," by D. A. Tyler. AM IND HYG ASSOC J 34:345-349, August, 1973.

"Noise and Uncle," by W. L. Clevenger. AUDIO ENG SOC J 21:724-726, November, 1973.

"Noise and vibration analysis of an impact forming machine," by A. E. M. Osman, et al. J ENG IND 96:233-240, February, 1974.

"Noise and vibration hazards," JAP J HYG 29:162-169, April, 1974.

"Noise and vibration of resiliently supported track slabs," by E. K. Bender. ACOUSTICAL SOC AM J 55:259-268, February, 1974.

"Noise and vibrations of agriculture tractors and measures of prevention," by D. Petrovi. NAR ZDRAV 29:110-115, April, 1973.

"Noise as a pollutant," by J. Connell. DIST NURS 15:196, December, 1972.

"Noise at the working place," by S. Pietruck, et al. ZENTRALBL ARBEITSMED 24:139-148, May, 1974.

"Noise considerations on large process unit drivers," by T. C. Again, et al. IEEE TRANS IND APPLICATIONS 10:296-304, March, 1974.

"Noise control act of 1972—congress acts to fill the gap in environmental legislation," MILL L REV 58:273-306, December, 1973.

"Noise control: an act where purchasing can take the lead," by C. H. Deutsch. PURCHASING 75:19 plus, December 18, 1973.

"Noise control and civil engineering," by E. M. Krokosky, et al. CIVIL ENG 44:45-49, May, 1974.

"Noise control: a common-sense approach," by R. L. Lowery. MECH ENG 95:26-31, June, 1973.

"Noise-control design for process plants," by S. C. Lou. CHEM ENG 80:77-82, November 26, 1973.

"Noise control does not have to be a problem: Moduline system." ENGINEER 239:22, July 4, 1974.

"Noise control enclosure improves dryer efficiency." ROADS & STS 116:142 plus, September, 1973.

"Noise control in Connecticut." DIESEL EQUIP SUPT 52:46, September, 1974.

"Noise control in Oregon: government regulation and private remedies," WILLAMETTE L J 10:198-216, Spring, 1974.

"Noise control leads to better bearing designs." PRODUCT ENG 45:21, January, 1974.

"Noise control office near action after years of study." IND W 181:18 plus, June 3, 1974.

"Noise control versus shock and vibration engineering," by C. T. Morrow. ACOUSTICAL SOC AM J 55:695-699, April, 1974.

"Noise coupling between accommodation and accommodative vergence," by D. Wilson. VISION RES 13:2505-2513, December, 1973.

"Noise environment of a typical school classroom due to the operation of utility helicopters," by D. A. Hilton, et al. J ACOUST SOC AM 55: Suppl:37, 1974.

"Noise exposure risks lessened: CEL Noise Dosimeter." ENGINEER 237:23, July 12, 1973.

"Noise factor in the main metallurgical production shops and measures for its control." by L. A. Sobolevskaia. GIG TR PROF ZABOL 16:3-7, July, 1972.

"Noise factors in product liability," by C. E. Wilson. QUALITY PROG

7:28-30, February, 1974.

"Noise from aerial bursts of fireworks," by D. J. Maglieri, et al. J ACOUST SOC AM 54:1224-1227, November, 1973.

"Noise from turbine drill. Risk for hearing injuries among dentists?" by T. Hundseth, et al. NOR TANNLAEGEFOREN TID 83:185-187, May, 1973.

"Noise: future targets," by G. M. Lilley. AERONAUTICAL J 78:459-463, October, 1974.

"Noise: the government's view," by R. E. Train. MED TIMES 102:63-64, March, 1974.

"A noise hazard to local authority employees," by W. A. Pollitt, et al. COMMUNITY HEALTH 5:19-23, July-August, 1973.

"Noise in hospital." (editorial). BR MED J 4:625, December 15, 1973.

"Noise in hospitals," by R. Taylor. HEALTH SOC SERV J 84:2770-2771, November 30, 1974.

"Noise in libraries: causes and control." SPECIAL LIBRARIES 65,1:28-31, January, 1974.

"Noise in the quiet zone." MOD HEALTHCARE, SHORT-TERM CARE ED 1:59-63, April, 1974.

"Noise-induced hearing loss," by F. J. Dittrich. J AM OSTEOPATH ASSOC 73:446-449, February, 1974.

"Noise-induced hearing loss and presbyacusis," by J. H. Macrae. AUDIOLOGY 10:323-333, September-December, 1971.

"Noise-induced hearing loss and snowmobiles," by F. H. Bess, et al. ARCH OTOLARYNGOL 99:45-51, January, 1974.

"Noise-induced hearing loss: the energy principle for recurrent impact noise and noise exposure close to the recommended limits," by G. R. Atherley. ANN OCCUP HYG 16(2):183-194, August, 1973.

"Noise-induced inner ear damage in newborn and adult guinea pigs," by S. A. Falk, et al. LAYRNGOSCOPE 84:444-453, March, 1974.

"Noise-induced reduction of inner-ear microphonic response: dependence on body temperature," by D. G. Drescher. SCIENCE 185:273-274, July 19, 1974.

"Noise-induced threshold shift in the parakeet (Melopsittacus undulatus)," by J. Saunders, et al. PROC NATL ACAD SCI USA 71:1962-1965, May, 1974.

"Noise is causing an industrial headache," by D. W. Austin. MED TIMES 102:60-63, March, 1974.

"Noise levels in infant incubators (adverse effects?)," by G. Blennow, et al. PEDIATRICS 53:29-32, January, 1974.

"Noise limitations on measurements made with SLUGS," by J. C. Gallop. SCI INSTR 7:855-859, October, 1974.

"Noise, man and law," by E. Hammelburg. ORL 35:363-370, 1973.

"Noise measurement and control," by R. K. Miller. SYSTEMS DESIGN 70:41-44, June, 1973.

"Noise measurement standards for machine in situ," by W. W. Lang. ACOUSTICAL SOC AM J 54:960-966, October, 1973.

"Noise measurement techniques," by J. Donovan. ASTM STAND N 2:17-31 plus, May, 1974.

"Noise measurements in a university: an open-ended student experiment," by A. A. Silvidi, et al. AMERICAN JOURNAL OF PHYSICS 41,7: 909-913, July, 1973.

"Noise: meter men move in." MEL MAKER 49:5, January 12, 1974.

"Noise! Noise! Noise!" by E. Kiester, Jr. FAMILY HEALTH 6:20-21, plus, January, 1974.

"Noise nuisance," by F. Holland. COUNTRY LIFE 156:630, September 5, 1974.

"Noise of industrial enterprises and its effect on the population of Krivoi Rog," by N. M. Paran'ko, et al. GIG SANIT 37:98-99, July, 1972.

"Noise; a pain in the ear." METALLURGIA & METAL FORMING 41:55, March, 1974.

"Noise pollution," (editorial), by P. W. Alberti. CAN J OTOLARYNGOL 1:279-280, 1972.

"Noise pollution in the engineering office," by R. E. Herzog. MACHINE DESIGN 45:66-71, July 26, 1973.

"Noise pollution in hospitals is regarded as health hazard." OR REPORTER 8:3 plus, October, 1973.

"Noise pollution in the woodworking laboratory," by C. A. Pinder. MAN/SOC/TECH 34:47-50, November, 1974.

"Noise pollution: just how bad is it?" by J. E. Watson. MED TIMES 102:51-59, March, 1974.

"A noise pollution level instrument," by J. A. Hamburg. REV SCI INSTRUM 44:1618-1620, November, 1973.

"Noise pollution: neonatal aspects." PEDIATRICS 54:476-479, October, 1974.

"Noise pollution: public needs vs. individual rights," by D. Kastan. WESTERN STATE L REV 185-216, June, 1974.

"Noise pollution seminar." MECH ENG 96:42, January, 1974.

"Noise propagation in cellular urban and industrial spaces," by H. G. Davies, et al. ACOUSTICAL SOC AM J 54:1565-1570, December, 1973.

"Noise-reducing punch-press guard," by R. S. Florczyk. PLANT ENG 27:158-159, October 18, 1973.

"Noise reduction by barriers," by U. J. Kurze. ACOUSTICAL SOC AM J 55:504-518, March, 1974.

"Noise studies and exposure tests in metallurgy," by L. Schreiner, et al.

ZENTRALBL ARBEITSMED 24:148-153, May, 1974.

"Noise, vitamin A deficiency, and emotional behavior in rats," by A. Wells, et al. PERCEPT MOT SKILLS 38:392-394, April, 1974.

" 'Noisy' battle over OSHA regulation," by H. G. Unger. CAN BUS 46:8, November, 1973.

"Noisy trains and noisy typewriters pose different acoustical problems: so get different acoustical treatment." ARCHIT REC 156:96-97, Mid-August, 1974.

"Nomograph determines effects of fan rpm on noise level: data sheet," by F. Caplan. HEATING-PIPING 45:49-50, December, 1973.

"Nonauditory of effects of noise," by M. Schiff. TRANS AM ACAD OPHTHALMOL OTOLARYNGOL 77:1384-1398, September-October, 1973.

"A non-verbal analogue to the verbal transformation effect," by N. J. Lass, et al. CAN J PSYCHOL 27:272-279, September, 1973.

"Not so silent service." HEALTH SOC SERV J 84:343, February 16, 1974.

"A note on the protection afforded by hearing protectors—implications of the energy principle," by D. Else. ANN OCCUP HYG 16:81-83, April, 1973.

"Occupational exposure to noise," by L. Hughes. J OCCUP MED 16:38, January, 1974.

"Occupational hygiene and health status of concrete workers in construction of large hydroelectric power plants," by G. N. Metlyaev. GIG TR PROF ZABOL 17(12):8-11, 1973.

"On the hearing of residents near airports," by D. C. Nagel, et al. J ACOUST SOC AM 55(2):463-464, 1974.

"On the noise level of ears and microphones," by M. C. Killion. J ACOUST SOC AM 55:Suppl:41, 1974.

"Opaque cisternography, using the sub-occipital route. (Method for the study of the cerebellopontile angle and the internal auditory canal. Indications and technic)," by V. Agnetti, et al. SIST NERV 23:6-19, 1971.

"Open-cell urethane foams offer pluses for noise control." PRODUCT ENG 45:37-38, April, 1974.

"Optimizing vent silencers design for gas blow-downs," by J. Hawkins. PIPELINES & GAS J 210:68 plus, October, 1974.

"The optimum human environment," by H. Hillman. NURS TIMES 69: 692-695, May 31, 1973.

"Oranges save Britain from rock." MACHINE DESIGN 45:47, December 13, 1973.

"OSHA, EPA disagree on operator exposure limits for noise," by E. Tabaczuk. AIR COND HEAT & REFRIG N 133:1 plus, December 9, 1974.

"OSHA issues tunnel rules, wrestles with noise." ENGIN N 192:10, April 18, 1974.

"OSHA noise: guilty! what did you say?" IRON AGE 212:36, November 22, 1973.

"OSHA noise standards too lenient, says EPA." CHEM MKTG REP 206:16, December 23, 1974.

"OSHA roundup: newspapers safe, noise problem mounts," by M. C. Fisk. ED & PUB 106:10 plus, December 15, 1973.

"Other medical views and research on noise." MED TIMES 102:84-86 plus, March, 1974.

"Otorhinolaryngology, scuba diving and hyperbaric medicine," by A. Appaix, et al. J FR OTORHINOLARYNGOL 22:559-561 plus, September, 1973.

"Outlook for in-situ measurement of noise from machines," by T. J. Schultz. ACOUSTICAL SOC AM J 54:982-984, October, 1973.

"An overview of EPA's implementation of the Noise Control Act of 1972," by A. F. Meyer, Jr. J AIR POLLUT CONTROL ASSOC 24:830-831, September, 1974.

"Owl's wing a noise inhibitor," by P. Soderman. MECH ENG 95:56, October, 1973.

"Participation in urban planning: the Barnsbury case [book review]," by R. Mordey. ROY TOWN PLAN INST J 59:2-3, January, 1974.

"Pathogenesis of occupational hearing disorder under the combined action of general vibration and noise," by I. P. Enin. ZH USHN NOS GORL BOLEZN 0(1):53-57, January-February, 1974.

"Patients blame staffs in hospital noise control report." NURSING TIMES 70:251, February 21, 1974.

"Performance during continuous and intermittent noise and wearing ear protection," by L. R. Hartley. J EXP PSYCHOL 102:512-516, March, 1974.

"Personal noise dosimetry in refinery and chemical plants," by A. H. Diserens. J OCCUP MED 16:255-257, April, 1974.

"Photographic analysis of human startle reaction to sonic booms," by R. I. Thackray, et al. AEROSP MED 45:803-806, August, 1974.

"Physiological correlates of auditory stimulus periodicity," by J. E. Hind. AUDIOLOGY 11:42-57, January-April, 1972.

"Physiological effects of intermittent noise," by Y. Osada, et al. J PHYSIOL SOC JAP 35:460, August-September, 1973.

"Physiological effects of noise. An overview," by B. L. Welch. FED PROC 32:2091-2092, November, 1973.

" 'Piano killer' strikes a responsive chord in noise-filled Japan: slaying of woman, two daughters over music sparks debate in an overcrowded nation," by N. Pearlstine. WALL ST J 184:1 plus, December 3, 1974.

"Pilatus develops quiet turbo Porter." AVIATION W 99:65-67, October 29, 1973.

"A pilot investigation of noise hazards in recording studios," by G. W. Gibbs, et al. ANN OCCUP HYG 16(4):321-327, 1973.

"Pinellas County's noise ordinance sets A/C sound levels at 60 dBA," by P. Kelsey. AIR COND HEAT & REFRIG N 133:3, November 11, 1974.

"Pitch and stimulus fine structure," by F. L. Wightman. J ACOUST SOC AM 54:397-406, August, 1973.

"Placement of electrodes for excitation of the eighth nerve," by R. A. Walloch, et al. ARCH OTOLARYNGOL 100:19-23, July, 1974.

"Plain film demonstration of acoustic nerve tumors," by L. E. Etter. ARCH OTOLARYNGOL 98:414-415, December, 1973.

"Planning and noise." (editorial). R SOC HEALTH J 93:58, April, 1973.

"Planning for airports in urban environments—a survey of the problem and its possible solutions," by M. L. Dworkin. TRANSP L J 5:183-214, July, 1973.

"Plastic tube silences machinery." MACHINE DESIGN 46:40, April 4, 1974.

"Platelet adhesiveness during noise exposure," by B. Maass, et al. DTSCH MED WOCHENSCHR 98:2153-2159, November 9, 1973.

"Polysensory interactions in the cuneate nucleus," by S. F. Atweh, et al. J PHYSIOL 238:343-355, April, 1974.

"Polysensory responses and sensory interaction in pulvinar and related postero-lateral thalamic nuclei in cat," by C. C. Huang, et al. ELECTROENCEPHALOGR CLIN NEUROPHYSIOL 34:265-280, March, 1073.

"Porous plastic silencers hush pneumatic presses." ENGINEER 237:23, October 18, 1973.

"Portable measuring device for the equivalent gauge of permanent noise," by W. Liebig. Z GESAMTE HYG 20:1-4, January, 1974.

"Positive progress announced on Illinois noise problem." HOT ROD 27:32,

June, 1974.

"Possibilities and problems of achieving community noise acceptance of VTOL," by W. Z. Stepniewski, et al. AERONAUTICAL J 77:311-326, June, 1973.

"A possible neurophysiological basis for the precedence effect," by I. C. Whitfield. RED PROC 33:1915-1916, August, 1974.

"Postnatal development of provoked activity in the superior lateral olive in the cat by stimulation by sound," by R. Romand, et al. J PHYSIOL 66:303-315, September, 1973.

"Practical problems in stopping on-the-job noise pollution," by R. D. Moran. J OCCUP MED 16:19-21, January, 1974.

"Predict plant noise problems," by R. S. Norman. HYDROCARBON PROCESS 52:89-91, October, 1973.

"Preliminary summary on the clinical experience with treatment of 77 cases of explosion deafness." CHIN MED J 4:238-241, 1974.

"Presbyacusis. VI. Masking of speech," by K. Jokinen. ACTA OTOLARYNGOL 76:426-430, December, 1973.

"Preventing hearing loss due to excessive noise exposure," by J. Sataloff. J OCCUP MED 16:470-471, July, 1974.

"Prevention of neurosensorial hypoacusia in metallurgic workers," by R. Benavides. REV MED CHIL 101:643-645, August, 1973.

"Priming for audiogenic seizures in mice: influence of postpriming auditory environment," by G. R. Bock, et al. EXP NEUROL 42:700-702, March, 1974.

"Problem of pseudosyringomyelitic ulcerative-mutilating and deforming acropathy," by P. I. Golemba, et al. VESTN DERMATOL VENEROL 47:73-75, August, 1973.

"The problem of traffic noise," by D. J. Fisk. R SOC HEALTH J 93:289-290 plus, December, 1973.

"Problems arising from studies concerning the convergence of noise-related hardness of hearing and hearing problems of other origins with special attention paid to agerelated hardness of hearing," by T. Brusis. Z LARYNGOL RHINOL OTOL 52:915-929, December, 1973.

"Production of calibrated sound pressures at the tympanic membrane of the guinea pig," by H. Wagner, et al. ARCH OTORHINOLARYNGOL 206:283-292, June 18, 1974.

"Program of hearing preservation in the metallurgy plant in Huachipato," by R. Benavides. REV MED CHIL 101:661-665, August, 1973.

"Property—eminent domain—even where no direct overflight occurs, aircraft noise and air pollution depriving property owners of the practical use and enjoyment of their land is a taking requiring compensation for the diminution of the land's market value." J URBAN L 52:636-648, Winter, 1974.

"Propulsion system design for the ATT," by G. L. Brines. J AIRCRAFT 10:487-490, August, 1973.

"Protection from lethal audiogenic seizures in mice by physical restraint of movement," by J. F. Willott. EXP NEUROL 43:359-368, May, 1974.

"Protection of residential areas from industrial noise," by P. Koltzsch, et al. Z GESAMTE HYG 19:331-337, May, 1973.

"Psychological effects of exposure to high industrial noise: a field study," by E. Gulian. J ACOUST SOC AM 55:Suppl:68, 1974.

"Public health nurse and the hearing damage," by J. Jensen. SYGEPLJERSKEN 72:6-7, July 13, 1972.

"Public health then and now. A backward glance at noise pollution," by G. Rosen. AM J PUBLIC HEALTH 64:514-517, May, 1974.

"Pure tone audiometric picture of noise induced deafness," by T. Brusis. Z LARYNGOL RHINOL OTOL 52:673-680, September, 1973.

"Putting all our noise technology to work," by R. P. Jackson. ASTRONAUTICS & AERONAUTICS 12:48-51, January, 1974; Discussion 12:4-6, April, 1974.

"Quantitative model for the effects of stimulus frequency upon synchronization of auditory nerve discharges," by D. J. Anderson. J ACOUST SOC AM 54:361-364, August, 1973.

"Quiet bagging; Vyon silencers aid in machinery noise reduction." COMP AIR MAG 79:8-9, July, 1974.

" 'Quiet campaign' featured at Methodist Hospital of Indianapolis, Ind." HOSP TOP 51:15, February, 1973.

"Quiet campaign—a sound idea." MICH HOSP 10:24, August, 1974.

"Quiet please. Noise does affect your health and well-being," by A. Magie. LIFE HEALTH 88:14-17, March, 1973.

"Quiet refuse truck beats 75 decibels." FLEET OWNER 69:170, March, 1974.

"Quiet truck program progressing at International." AUTOMATIVE IND 150:138 plus, April 1, 1974.

"Quieted air cylinders meet OSHA noise requirements." HYDRAULICS & PNEUMATICS 26:133, September, 1973.

"Quieter please!" by I. Berkovitch. ENGINEERING 214:392-394, May, 1974.

"Quieting of process machinery," by H. A. Winnerling. CHEM ENG PROG 69:96-99, June, 1973.

"Quietness as a remedy," by Z. Naumowski. PIELEG POLOZNA 4:23-24, April, 1973.

"Radiation therapy of tumors of the eighth nerve sheath," by H. Newman, et al. AM J ROENTGENOL RADIUM THER NUCL MED 120:562-567, March, 1974.

"Radio noise of terrestrial origin; abstracts of papers." RADIO SCI 8:613-621, June, 1973.

"Ralph Nader reports," by R. Nader. LADIES HOME J 91:22 plus, January, 1974.

"Rebuttal to papers on aircraft noise by S. R. Lane at the 86th Meeting of the ASA," by V. E. Callaway. J ACOUST SOC AM 55:1343-1345, June, 1974.

"Recent physical examination technics of speech," (proceedings), by F. J. Landwehr, et al. ARCH KLIN EXP OHREN NASEN KEHLKOPF-HEILKD 205:388-391, December 17, 1973.

"Reception of consonants in a classroom as affected by monaural and binaural listening, noise, reverberation, and hearing aids," by A. K. Nabelek, et al. J ACOUST SOC AM 56(2):628-639, August, 1974.

"The recovery cycle of the averaged auditory evoked response during sleep in normal children," by E. M. Ornitz, et al. ELECTROENCEPHALOGR CLIN NEUROPHYSIOL 37:113-122, August, 1974.

"Recovery from sound exposure in auditory-nerve fibers," by E. Young, et al. J ACOUST SOC AM 54:1535-1543, December, 1973.

"Recovery of detection probability following sound exposure: comparison of physiology and psychophysics," by E. Young, et al. J ACOUST SOC AM 54:1544-1553, December, 1973.

"Recreational noise: implications for potential hearing loss to participants," by J. H. Shirreffs. J SCH HEALTH 44:548-550, December, 1974.

"Reducing noise in food plants," by R. K. Miller. FOOD ENG 46:75-76, February, 1974.

"Reduction of VTOL operational noise through flight trajectory management," by F. H. Schmitz, et al. J AIRCRAFT 10:385-394, July, 1973.

"The relation between temporary threshold shift and permanent threshold shift in rhesus monkeys exposed to impulse noise," by G. A. Luz, et al. ACTA OTOLARYNGOL [Suppl] 1-15, 1973.

"Relation of noise measurements to temporary threshold shift in snowmobile users," by R. B. Chaney, Jr., et al. J ACOUST SOC AM 54: 1219-1223, November, 1973.

"Relations between the psychophysics and the neurophysiology of sound localization," by G. Moushegian, et al. FED PROC 33:1924-1927, August, 1974.

"The relationship between the length of exposure to noise and the incidence of hypertension at a silo in Terran," by N. Kavoussi. MED LAV 64:292-295, July-August, 1973.

"The relationship between permanent threshold shift and the loss of hair cells in monkeys exposed to impulse noise," by M. Pinheiro, et al. ACTA OTOLARYNGOL [Suppl] 31-40, 1973.

"Reply to criticisms by V. E. Callaway of papers MM1 and MM11 at the 86th meeting of the ASA," by S. R. Lane. J ACOUST SOC AM 55: 1346-1348, June, 1974.

"Research on the unitary discharges of cells of the anterior sigmoid gyrus after acoustic stimulation in normal animals and in animals whose acoustic areas have been removed bilaterally," by O. Sager, et al. REV ROUM NEUROL 10:61-73, 1973.

"Resilient hub hushes fan." MACHINE DESIGN 45:35, December 27, 1973.

"Response of the amphibian papilla nerve in the toad bufomarinus," by H. Oyama. J PHYSIOL SOC JAP 35:534-535, August-September, 1973.

"The response of the swim bladder of the goldfish (Carassius auratus) to acoustic stimuli," by A. N. Popper. J EXP BIOL 60:295-304, April, 1974.

"Responses of neurons in auditory cortex of the macaque monkey to monaural and binaural stimulation," by J. F. Brugge, et al. J NEUROPHYSIOL 36:1138-1158, November, 1973.

"Responsiveness to simple and complex auditory stimuli in the human newborn," by G. Turkewitz, et al. DEV PSYCHOBIOL 5:7-19, 1972.

"Reverberation in a city street," by D. Aylor, et al. ACOUSTICAL SOC AM J 54:1754-1757, December, 1973.

"Rhythmic structure in auditory temporal pattern perception and immediate memory," by P. T. Sturges, et al. J EXP PSYCHOL 102:377-383, March, 1974.

"Role of auditory cortex in sound localization: a comparative ablation study of hedgehog and bushbaby," by R. Ravizza, et al. FED PROC 33:1917-1919, August, 1974.

"Role of centrifugal pathways to cochlear nucleus in detection of signals in noise," by J. O. Pickles, et al. J NEUROPHYSIOL 36:1131-1137, November, 1973.

"Role of the fornix in the septal syndrome," by D. S. Olton, et al. PHYSIOL BEHAV 13(2):269-279, August, 1974.

"The role of green plants in the prevention of air and noise pollution," by G. Zamfir. REV MED CHIR SOC MED NAT IASI 77:673-678, October-December, 1973.

"Role of multiple reflections and reverberation in urban noise propagation," by R. H. Lyon. ACOUSTICAL SOC AM J 55:493-503, March, 1974.

"Role of the personnel department in a hearing conservation program," by B. Harmon. PERS J 53:531-535, July, 1974.

"Role of sound, transmitted through the air or a substrate, in the communications of social insects," by E. K. Es'kov. ZH OBSHCH BIOL 34: 861-871, November-December, 1973.

"Rubber pad quiets wheel squeal." MACHINE DESIGN 46:42, September 5, 1974.

"School children in London; noise survey." HEALTH VISIT 45:174-277, September, 1972.

"Sensitization of the rat startle response by noise," by M. Davis. J COMP PHYSIOL PSYCHOL 87(3):571-581, September, 1974.

"Shielding cuts truck and bus noise." AUTOMOTIVE ENG 81:15-16, August, 1973.

"Shock tolerance in rats as a function of white noise," by M. Cunningham, et al. PSYCHOL REP 34:711-713, June, 1974.

"Should the practice for financial compensation for occupational acoustic trauma be changed?" by A. Ahlmark, et al. LAKARTIDNINGEN 70: 3151-3154, September 12, 1973.

"Should you colour your [office] sound?" by L. W. Hegvold. OPTIMUM 5,1:54-60, 1974.

"Shut up those sudden noises." ENGINEER 237:23, November 29, 1973.

"Signal-to-noise ratio as a predictor of startle amplitude and habituation in the rat," by M. Davis. J COMP PHYSIOL PSYCHOL 86:812-825, May, 1974.

"Silencer for jet whine." NEWSWEEK 84:34, July 1, 1974.

"Silencing of generators." ENGINEER 238-18, February 28, 1974.

"Silencing the snowmobile." AUTOMOTIVE ENG 81:72-73 plus, September, 1973.

"Silent air; Atlas Copco Silensair STS 71." ENGINEERING 213:425, June, 1973.

"Silent cabins, silencers, machine enclosures." ENGINEER 239:18, September 12, 1974.

"Singing muscles in a katydid," (letter), by J. W. Pringle. NATURE 250: 442, August 2, 1974.

"Slow cortical evoked potentials: interactions of auditory, vibro-tactile and shock stimuli," by I. S. Hay, et al. AUDIOLOGY 10:9-17, January-February, 1971.

"Social impact of aircraft noise," by A. Alexandre. TRAFFIC Q 28:371-388, July, 1974.

"The 'soft-spoken' woman. II. Auditory vs non-auditory monitoring of loudness behaviors," by P. J. Dembowski, et al. J COMMUN DISORD 6:206-212, September, 1973.

"Solution of the noise problem with the Aue II artificial kidney," by J. Iversen, et al. Z UROL NEPHROL 66:919-920, December, 1973.

"Some aspects of the problem of adaptation to noise," by E. Ts. Andreeva-Galanina, et al. GIG SANIT 38:34-37, December, 1973.

"Some effects of noise on the speaking behavior of stutterers," by E. G. Conture. J SPEECH & HEARING RES 17:714-723, December, 1974.

"Some experiments relating to the perception of complex tones," by B. C. Moore. Q J EXP PSYCHOL 25:451-475, November, 1973.

"Some implications regarding high frequency hearing loss in school-age children," by R. L. Cozad, et al. J SCH HEALTH 44:92-96, February, 1974.

"Some methodological problems in studying the action of noise on the body of man and animals," by L. I. Maksimova, et al. GIG SANIT 38: 30-35, July, 1973.

"Some observations on the effects of attention to stimuli on the amplitude of the acoustically evoked response," by L. W. Keating, et al. AUDIOLOGY 10:177-184, May-June, 1971.

"Some techniques for assessment of community noise environments," by L. C. Sutherland, et al. J ACOUST SOC AM 55(2):464, 1974.

"Sound advice." (editorial). NURS TIMES 70:249, February 21, 1974.

"Sound and sound emission apparatus in puerulus and postpuerulus of the western rock lobster (Panulirus longipes)," by V. B. Meyer-Rochow, et al. J EXP ZOOL 189:283-289, August, 1974.

"Sound deprivation causes hypertension in rats," by M. F. Lockett, et al. FED PROC 32:2111-2114, November, 1973.

"Sound detector of X-ray and gamma-radiation," by T. V. Lapchuk, et al. MED RADIOL 19(7):77-82, July, 1974,

"Sound intensity and good health," by H. Haggerty. PHYS TEACH 12:421-423, October, 1974.

"Sound-level measurements in the community," by D. S. Allen. AIR COND HEAT & REFRIG N 131:30-32, March 4, 1974.

"Sound levels recommended for mobile construction equipment." AUTOMOTIVE ENG 81:23, June, 1973.

"Sound pollution," by R. Moody. SCI & CHILD 12:6, October, 1974.

"Sound power measurements on large machinery installed indoors," by G. M. Diehl. COMP AIR MAG 79:8-12, January, 1974.

"Sound-power testing experiences of an independent laboratory," by M. J. Kodaras, et al. ACOUSTICAL SOC AM J 54:956-959, October, 1973.

"Sound production in scolytidae: 'rivalry' behaviour of male Dendroctonus beetles," by J. A. Rudinsky, et al. J INSECT PHYSIOL 20:1219-1230, July, 1974.

"Sound spectrographic examinations of heart sounds and murmurs in aortic valve lesions," by A. Aigner, et al. Z KARDIOL 63(3):269-278, March, 1974.

"Sound-technical viewpoints for the planning and realization of heating centers," by H. Schmitz. GESUND ING 92:141-146, May, 1971.

"Sound wave receiving mechanism seen from the aspect of comparative physiology," by Y. Katsuki. J OTOLARYNGOL JAP 76:1297-1300, October, 1973.

"Soundproofing your engine compartment," by J. Martenhoff. MOTOR B & S 134:67-69, July, 1974.

"Sounds around us." AIR COND HEAT & REFRIG N 133:39, November 4, 1974.

"Sources of noise and lowering of the noise level in the dental office," by W. Hoefig. DTSCH ZAHNAERZTL Z 99:172-175, February, 1973.

"Spaced can feeding reduces noise from 100 to 85 db [Shasta]," by E. Lane, et al. FOOD PROCESSING 34:38-39, November, 1973.

"Spatial stimulus generalization as a function of white noise and activation level," by R. E. Thayer, et al. J EXP PSYCHOL 102:539-542, March, 1974.

"Speakers and rooms," by R. Hodges. POP ELECTR 6:22-26, August, 1974.

"Specific acute losses of vestibular function in four patients following unilateral section of one or all components of the eighth cranial nerve," by E. F. Miller II, et al. ANN OTOL RHINOL LARYNGOL 83:525-536, July-August, 1974.

"Spectrographic analysis of fundamental frequency and hoarseness before and after vocal rehabilitation," by M. Cooper. J SPEECH HEAR DISORD 39(3):286-297, August, 1974.

"Speech acoustics for the theatre," by D. L. Klepper. AUDIO ENG SOC J 22:15-19, January-February, 1974.

"Sports, noisy and quiet," (letter), by R. H. Nuenke. N ENGL J MED 290:523, February 28, 1974.

"Square-wave stimuli and neonatal auditory behavior: reply to Bench," by S. J. Hutt. J EXP CHILD PSYCHOL 16:530-533, December, 1973.

" 'Square-wave stimuli' and neonatal auditory behavior: some comments on Ashton (1971), Hutt et al. (1968) and Lenard et al. (1969)," by J. Bench. J EXP CHILD PSYCHOL 16:521-527, December, 1973.

"SST noise levels called acceptable." AVIATION W 100:26, March 18, 1974.

"State of the acoustic analyzer in stuttering," by A. G. Rakhmilevich. ZH USHN NOS GORL BOLEZN 33:27-31, September-October, 1973.

"State of auditory function in persons working since adolescence in a noisy environment," by I. B. Kramarenko, et al. VESTN OTORINOLARINGOL 35:93-96, January-February, 1973.

"State regulation of nontransportation noise: law and technology," by R. W. Findley, et al. SO CALIF L REV 48:209-317, November, 1974.

"State standards, regulations, and responsibilities in noise pollution control," by J. M. Tyler, et al. J AIR POLLUT CONTROL ASSOC 24:130-135, February, 1974.

"Statistical analysis of continuous data records," by R. B. Corotis. AM SOC C E PROC 100:195-206, February, 1974.

"Statistical analysis of telephone noise," by B. W. Stuck, et al. SYSTEM TECH J 53:1263-1320, September, 1974.

"Stimulation interval-dependent acoustic evoked potentials," by E. Sturzebecher, et al. ACTA OTOLARYNGOL 77:256-260, April, 1974.

"Stimulus intensity and recency contrasts and orienting response strength," by D. C. Edwards. PSYCHOPHYSIOLOGY 11(5):543-547, September, 1974.

"Stop noise with hedges," by C. E. Whitcomb, et al. HORTICULTURE 52: 58-59, April, 1974.

"Stop plant noise at the source or along the way," by C. L. Meteer. AUTOMATION 21:58-61, July, 1974.

"Stopping hydraulic system noise," by H. W. Wojda. PLANT ENG 27:74-75, July 26, 1973.

"Stria ultrastructure and vessel transport in acoustic trauma," by A. J. Duvall III, et al. ANN OTOL RHINOL LARYNGOL 83:498-514, July-August, 1974.

"Stronger rules recommended to cut propeller aircraft noise." AVIATION W 100:61, March 25, 1974.

"Studies on the diagnostic significance of the galvanic test in acoustic neurinoma," by C. R. Pfaltz. ARCH KLIN EXP OHREN NASEN KEHLKOPFHEILKD 205:130-134, December 17, 1973.

"Studies on noise in small cities. 2. City noise in Karatsu City," by S. Sato, et al. JAP J HYG 28:425-428, October, 1973.

"Studies on noise in small cities. 3. Community reaction to noise in Karatsu City," by H. Miura, et al. JAP J HYG 28:429-436, October, 1973.

"Studies on the perception of acoustic signals under noise-protection conditions in track-packing work, and corresponding conclusions," by K. Jungsbluth, et al. ZENTRALBL ARBEITSMED 24:153-156, May, 1974.

"Study of cerebral circulation under the separate and joint action of intensive noise and physical load," by I. B. Evdokimova, et al. GIG TR PROF ZABOL 17:1-5, July, 1973.

"A study of laser-acoustic air pollution monitors," by L. G. Rosengren, et al. J PHYS E SCI INSTRUM 7:125-133, February, 1974.

"The subtemporal transtentorial approach for large acoustic nerve tumors," by F. Garcia-Bengochea, et al. ACTA NEUROL LAT AM 18:344-354, October-December, 1972.

"Superior vestibular and 'singular nerve' section—animal and clinical studies," by H. Silverstein, et al. LARYNGOSCOPE 83:1414-1432, September, 1973.

"Supervisor overseas: hospital noise," by J. Wakeley. SUPERV NURSE 5: 50, June, 1974.

"Surface transportation noise and its control," by J. E. Wesler. AIR POLLUTION CONTROL ASSN J 23:701-703, August, 1973.

"Susceptibility to damage from impulse noise: chinchilla versus man or monkey," (letter), by G. A. Luz, et al. J ACOUST SOC AM 54:1750-1754, December, 1973.

"Symposium: new data for noise standards. I. New data for noise standards," by D. Henderson, et al. LARYNGOSCOPE 84:714-721, May, 1974.

"Symposium: new data for noise standards. IV. The physiological effects of priming for audiogenic seizures in mice," by J. C. Saunders. LARYNGOSCOPE 84:750-756, May, 1974.

"Symptomatology, therapy and prognosis of acoustic neurinoma (a critical report on 118 cases)," by U. Borchardt, et al. PSYCHIATR NEUROL MED PSYCHOL 25:472-487, August, 1973.

"The symptoms of neuro-fibroma of the 8th nerve," by C. Ozsahinoglu, et al. J LARYNGOL OTOL 88:493-502, June, 1974.

"Synaptic transmission from hair cells to auditory fibers," (proceedings), by T. Furukawa, et al. J PHYSIOL SOC JAP 35:534, August-September, 1973.

"Synthesis of low-frequency noise for use in biological experiments," by A. S. French. IEEE TRANS BIOMED ENG 21:251-252, May, 1974.

"Tails of tuning curves of auditory-nerve fibers," by N. Y. Kiang,et al. J ACOUST SOC AM 55:620-630, March, 1974.

"Takahashi's work on jet noise," by S. Kondo, et al. J HUM ERGOL 2: 93-94, September, 1973.

"Take the thunder out of the big rigs: mandate from the EPA," by F. E. Bryson. MACHINE DESIGN 46:24-26 plus, September 19, 1974.

"Take the whine out of drawtwisting, and boost yarn quality as a bonus." TEXTILE WORLD 123:38 plus, November, 1973.

"Tale of two noise-suppression treatments." MACHINE DESIGN 45:12, November 1, 1973.

"Technique and value of gas and Pantopaque cisternography in the diagnosis of cerebello-pontine angle tumours," by S. Wende, et al. NEURORADIOLOGY 2:24-29, March, 1971.

"A technique for individual noise exposure assessment," by R. Colman, et al. J ACOUST SOC AM 55(2):464, 1974.

"Techniques for industrial noise measurement," by R. A. Boole. PLANT ENG 28:105-107, February 7, 1974.

"Temporary hearing losses in teenagers attending repeated rock-and-roll sessions," by R. F. Ulrich. ACTA OTO-LARYNGOL 77(1-2):51-55, 1974.

"Temporary threshold shift from a toy cap gun," by L. Marshall, et al. J SPEECH HEAR DISORD 39:163-168, May, 1974.

"Test and evaluation of a quiet helicopter configuration HH-43B," by M. A. Bowes. ACOUSTICAL SOC AM J 54:1214-1218, November, 1973.

"Theoretical prediction of highway noise fluctuations," by A. H. Marcus. J ACOUST SOC AM 56:132-136, July, 1974.

"Theory of binaural interaction based on auditory-nerve data. I. General strategy and preliminary results on interaural discrimination," by H. S. Colburn. J ACOUST SOC AM 54:1458-1470, December, 1973.

"Therapy of patients with acoustic trauma," by G. A. Dokukina, et al. VOEN MED ZH 8:35-38, August, 1973.

"Threshold shifts produced by exposure to noise in chinchillas with noise-induced hearing losses," by J. H. Mills. J SPEECH HEAR RES 16: 700-708, December, 1973.

"Thrust reverser noise estimation," by M. R. Fink. J AIRCRAFT 10:507-508, August, 1973.

"TLVs threshold limit values for physical agents adopted by ACGIH for 1973." J OCCUP MED 16:49-58, January, 1974.

"TMJ sound prints: electronic auscultation and sonagraphic audiospectral analysis of the temporomandibular joint," by P. L. Ouellette. J AM DENT ASSOC 89(3):623-628, September, 1974.

"Total environmental noise problem," by A. Glorig. ASSE J 19:22-26, July, 1974.

"Toward the comprehensive abatement of noise pollution: recent federal and New York city noise control legislation." ECOLOGY L Q 4:109-144, Winter, 1974.

"Traffic noise and overheating in offices." BUILD RES ESTAB DIGEST 162:1-4, February, 1974.

"Transformation function of the external ear in response to impulsive stimulation," by G. R. Price. J ACOUST SOC AM 56:190-194, July, 1974.

"Truck noise problem and what might be done about it," by R. F. Ringham. AUTOMOTIVE ENG 81:29-31 plus, April, 1973.

"Trucks are going to be quieter." IND W 181:57, April 8, 1974.

"Ulcero-mutilating acroplathies and diabetic perforating plantar disease," (letter), by C. Schmidt, et al. NOUV PRESSE MED 3:528, March 2, 1974.

"Ultrasound perception in pathologic processes of the cerebello-pontine angle area," by N. S. Biagoveshchensiaia, et al. VOPR NEIROKLIR 37:18-22, September-October, 1973.

"Ultrastructure of the spiral organ of the cochlea under normal conditions and following an experimental acoustic trauma," by O. Sh. Goniashvili. VESTN OTORINOLARINGOL 35:58-63, September-October, 1973.

"Understanding decibels," by G. Board. POP ELECTR 5:94-95, April, 1974.

"Understanding how noise affects hearing." FOUNDRY 101:79-80 plus, July, 1973.

"Undiagnosed acoustic neurinomas. A presentation of 4 cases," by J. Thomsen, et al. ARCH KLIN EXP OHREN NASEN KEHIKOPFHEILKD 204:175-182, October 12, 1973.

"Unexpected discovery of 2 neurinomas of the 8th pair in transtemporal surgery," by R. De Asis Alonso, et al. ACTA OTORINOLARYNGOL IBER AM 23:840-848, 1972.

"Unified analysis of fan stator noise," by D. B. Hanson. ACOUSTICAL SOC AM J 54:1571-1591, December, 1973.

"Unified approach to aerodynamic sound generation in the presence of solid boundaries," by M. Goldstein. ACOUSTICAL SOC AM J 56:497-509, August, 1974.

"Unitary mfrs detail price, time penalties of quieter equipment." AIR COND & REFRIG N 130:27, December 3, 1973.

"U.S. noise standards." LABOUR GAZ 74:692, October, 1974.

"U.S. officials find Concorde noise acceptable." AVIATION W 100:27, June 24, 1974.

"Unobtrusive sound reinforcement for an open-plan school." ARCHIT REC 156:151-152, September, 1974.

"An unusual acoustic neurinoma localized between brain stem and basilar artery using emulsified pantopaque cisternography," by J. L. Fox, et al. SURG NEUROL 2:329-332, September, 1974.

"Urban environment: noise and transportation. Environmental backlash—the urban paradox, noise and transportation, by A. G. Greenwald; Regulation-local, state and federal, by J. V. Tunner; Standards and controls, by A. F. Meyer, Jr.; Compliance and technology, by W. J. K. Gibson; Rights, remedies and planning, by D. C. McGrath, Jr.," NATURAL RESOURCES LAW 7:293-323, Spring, 1974.

"The use of alternated stimuli to reduce response decrement in the auditory testing of newborn infants," by D. Ling, et al. J SPEECH HEAR RES 14:531-534, September, 1971.

"Use of Xavin in auditory nerve lesions," by S. Spellenberg. THER HUNG 19:53-58, 1971.

"The validity of the 'energy principle' for noise-induced hearing loss," by H. Scheiblechner. AUDIOLOGY 13:93-111, March-April, 1974.

"The value of directional audiometry in the assessment of the central components in noise induced hearing loss," by H. G. Dieroff. Z LARYNGOL RHINOL OTOL 52:681-686, September, 1973.

"Value of foot dearterialization in ulceromutilating acropathies," (letter), by C. Lefaucher, et al. NOUV PRESSE MED 2:2958, December 8, 1973.

"Variables influencing performance on speech-sound discrimination tests," by A. H. Schwartz, et al. J SPEECH & HEARING RES 17:25-32, March, 1974.

"Vehicular noise regulation in Hawaii," by J. C. Burgess. ACOUSTICAL SOC AM J 56:905-910, September, 1974.

"Vestibular and auditory cortical projection in the guinea pig (Cavia porcellus)," by L. M. Odkvist, et al. EXP BRAIN RES 18:279-286, October 26, 1973.

"Vestibular apparatus and occupational deafness," by P. Picar. ACTA OTORHINOLARYNGOL BELG 26:657-663, 1972.

"Vestibular disturbances in the early diagnosis of tumors of the acoustic nerve," by L. B. Jongkees. J FR OTORHINOLARYNGOL 22:787-794, November, 1973.

"VFW 614; the quietest jet transport in the world." AIRCRAFT ENG 46:5, June, 1974.

"Vibration and noise in piping systems," by S. J. Shuey, et al. POWER ENG 77:42-44, June, 1973.

"Vibrations during construction operations," by J. F. Wiss. AM SOC C E PROC 100:239-246, September, 1974.

"Voltage noise, current noise and impedance in space clamped squid giant acon," by E. Wanke, et al. FFLUEGERS ARCH 347:63-74, February 18, 1974.

"Voluntary noise pacts likely to spread." IND W 182:23-25, August 19, 1974.

"Volvo designs a quiet tractor cab." AUTOMOTIVE ENG 81:68-71, September, 1973.

"Welding and ear injuries," by L. Andreasson, et al. LAKARTIDNINGEN 71:2443-2554, June 19, 1974.

"What is noise?" SCI DIGEST 75:62, May, 1974.

"What you must do about controlling noise." MOD MATERIALS HANDLING 29:44-49, February, 1974.

"What you ought to know and do about reducing noise in the pressroom," by W. H. Rouse. PTR/AM LITH 172:69-70 plus, April, 1974.

"What's behind the proposed truck noise regulations?" AUTOMOTIVE ENG 82:42-45 plus, July, 1974.

"What's so important about acoustics?" by A. Tipton. SCH MUS 45:18 plus, May, 1974.

"What's to hear," by C. Hopkins. J ARKANSAS MED SOC 71:87-98, July, 1974.

"Why noise reduction doesn't always work." MACHINE DESIGN 46:132, May 2, 1974.

"Why we're disturbed about noise," by C. A. Ragan, Jr. MED TIMES 102: 17 plus, March, 1974.

"Widened practice for compensation justified—temporary solutions must be possible to create," by A. Ahlmark, et al. LAKARTIDNINGEN 71:823, February 27, 1974.

"Workers' choice... ear defenders." OCCUP HEALTH 26:474, December, 1974.

"You can learn to live with noise control," by W. A. Bradley. POWER 117: 66-67, September, 1973.

PERIODICAL LITERATURE

SUBJECT INDEX

ACOUSTIC NERVE
see also: Noise Research

"Acoustic neurinoma. A comparison of the clinical picture and the electroencephalogram," by Z. Mensikova, et al. SB VED PR LEK FAK KARLOVY UNIV 15:401-409, 1972.

"Acoustic neuroma in the last months of pregnancy," by J. Allen, et al. AM J OBSTET GYNECOL 119:516-520, June 15, 1974.

"Acoustic neurinoma presenting as subarachnoid hemorrhage. Case report," by K. McCoyd, et al. J NEUROSURG 41(3):391-393, September, 1974.

"Acoustic neurinomas presenting as middle ear tumors," by L. A. Storrs. LARYNGOSCOPE 84:1175-1180, July, 1974.

"The acoustic reflex in eighth nerve disorders," by J. Jerger, et al. ARCH OTOLARYNGOL 99:409-413, June, 1974.

"Adrenal insufficiency and electrophysiological measures of auditory sensitivity," by F. W. Conn, et al. AM J PHYSIOL 225:1430-1436, December, 1973.

"Analysis of central nervous system involvement in the microwave auditory effect," by E. M. Taylor, et al. BRAIN RES 74:201-208, July 12, 1974.

"Angiographic diagnosis of acoustic neurinomas: analysis of 30 lesions," by M. Takahaski, et al. NEURORADIOLOGY 2:191-200, September, 1971.

"Audiologic evaluation in cochlear and eighth nerve disorders," by J. W. Sanders, et al. ARCH OTOLARYNGOL 100(4):283-289, October, 1974.

"Audiological comparison of cochlear and eighth nerve disorders," by J. Jerger, et al. ANN OTOL RHINOL LARYNGOL 83:279-285, May-June, 1974.

"Auditory information processing of vocalization," (proceedings), by K. Murata, et al. J PHYSIOL SOC JAP 35:535, August-September, 1973.

"Biopotentials of the organ of hearing in chronic sodium fluoride poisoning," by M. Kowalewska. OTOLARYNGOL POL 28(4):417-424, 1974.

"Brief-tone audiometry with normal, cochlear, and eighth nerve tumor patients," by W. O. Olsen, et al. ARCH OTOLARYNGOL 99:185-189, March, 1974.

"Cerebral vascular accidents in the course of tumors of the cerebellopontine angle. Pathogenic considerations," by C. Arseni, et al. EUR NEUROL 10:144-159, 1973.

"Chronic intracochlear electrode implantation: cochlear pathology and acoustic nerve survival," by R. A. Schindler, et al. ANN OTOL RHINOL LARYNGOL 83:202-215, March-April, 1974.

"Cochlear findings in 8th nerve tumors," by S. Katinsky, et al. AUDIOLOGY 11:213-217, May-August, 1972.

"Cochlear neurons: frequency selectivity altered by perilymph removal," by D. Robertson. SCIENCE 186:153-155, October 11, 1974.

"The co-existence of acoustic neuroma and otosclerosis," by J. D. Clemis. LARYNGOSCOPE 83:1959-1985, December, 1973.

"Consequences of peripheral frequency selectivity for nonsimultaneous masking," by H. Duifhuis. J ACOUST SOC AM 54:1471-1488, December, 1973;

"Contrast medium studies of the internal acoustic meatus (cisternomeatography)—a system for early diagnosis of acoustic nerve tumors," by I. Fleszar. POL PRZEGL RADIOL 38:9-17, 1974.

"Diagnosis of non-tumorous lesion of the cerebello-pontile region: with special reference to the differential diagnosis of acoustic nerve tumor," by Y. Yoshimoto. JAP J CLIN MED 31:3251-3260, November, 1973.

"Diagnostic value of Bekesy comfortable loudness tracings," by J. Jerger, et al. ARCH OTOLARYNGOL 99:351-360, May, 1974.

"Differential diagnosis of cerebello-pontine tumours," by J. Helms. LARYNGOL RHINOL OTOL 53:194-199, March, 1974.

"Differential phylogenetic development of the acoustic nuclei among chiroptera," by G. Baron. BRAIN BEHAV EVOL 9:7-40, 1974.

"Early diagnosis and management of acoustic neuromas," by J. M. Tew, Jr., et al. OHIO STATE MED J 70:365-367, June, 1974.

"Early diagnosis of neurinoma of the acoustic nerve (typical and atypical forms)," by J. M. Sterkers. PROBL ACTUELS OTORHINOLARYNGOL 29-41, 1971.

"Electrocochleography," by F. B. Simmons. ANN OTOL RHINOL LARYNGOL 83:312-313, May-June, 1974.

"Exposure of the internal auditory meatus by the House-Fisch-Portmann method with transsection of part of the 8th nerve," by B. Latkowski, et al. OTOLARYNGOL POL 27:569-575, 1973.

"Functional and histological findings in acoustic tumor," by M. Igarashi, et al. ARCH OTOLARYNGOL 99:379-384, May, 1974.

"Functional state of the auditory analyzer under conditions of prolonged clinostatic hypokinesia," by Z. I. Matsnev. VOEN MED ZH 7:62-65, July, 1973.

"Functional tests in lesions of the trigeminal n (V), acoustic n (V3) and the nervus intermedius in the region of the cerebello-pontine angle after Dandy's operation," by U. Koch, et al. Z LARYNGOL RHINOL OTOL 52:729-736, October, 1973.

"Further data, contributed by electrocochleography, on the function of normal and pathological peripheral receptors," by J. M. Aran. ACTA OTORHINOLARYNGOL BELG 26:671-683, 1972.

"Gammargraphic diagnosis in tumors of the 8th pair," by J. M. Sampere, et al. ARCH NEUROBIOL 37:45-60, January-February, 1974.

"Hitselberger' sign- its significance in the diagnosis of acoustic neurinoma," by H. Weidauer, et al. ARCH KLIN EXP OHREN NASEN KEHLKOPFHEILKD 205:126-130, December 17, 1973.

"Intact vestibular and cochlear function in acoustic neuroma," by W. S. Gunasekera, et al. CEYLON MED J 18:113-115, June, 1973.

"Intracellular septate desmosome-like structures in a human acoustic Schwannoma in vitro," by F. K. Conley, et al. J NEUROCYTOL 2:457-464, December, 1973.

"Isotopic and neuroradiologic correlation in the examination of auditory nerve neurinoma," by C. Bamberger-Bozo, et al. ROENTGENBLAETTER 26:182-189, April, 1973.

"Labyrinthectomy: indications, technique and results," by J. L. Pulec. LARYNGOSCOPE 84(9):1552-1573, September, 1974.

"Microsurgery of the internal auditory canal," by J. M. Sterkers. PROBL ACTUELS OTORHINOLARYNGOL 75-88, 1972.

"Middle ear measurements," by G. T. Wolcott, et al. J MED ASSOC STATE ALA 43:496-498, February, 1974.

"Model for mechanical to neural transduction in the auditory receptor," by M. R. Schroeder, et al. J ACOUST SOC AM 55:1055-1060, May, 1974.

"Neurilemmoma of the ninth cranial nerve masquerading as an acoustic neuroma, by J. R. Mountjoy, et al. ARCH OTOLARYNGOL 100: 65-67, July, 1974.

"Neurinoma of the acoustic nerve," by G. Djupesland. TIDSSKR NOR LAEGEFOREN 93:1751-1753, September 10, 1973.

"Opaque cisternography, using the sub-occipital route. (Method for the study of the cerebellopontile angle and the internal auditory canal. Indications and technic)," by V. Agnetti, et al. SIST NERV 23:6-19, 1971.

"Physiological correlates of auditory stimulus periodicity," by J. E. Hind. AUDIOLOGY 11:42-57, January-April, 1972.

"Placement of electrodes for excitation of the eighth nerve," by R. A. Walloch, et al. ARCH OTOLARYNGOL 100:19-23, July, 1974.

"Plain film demonstration of acoustic nerve tumors," by L. E. Etter. ARCH OTOLARYNGOL 98:414-415, December, 1973.

"Problem of pseudosyringomyelitic ulcerative-mutilating and deforming acropathy," by P. I. Golemba, et al. VESTN DERMATOL VENEROL 47:73-75, August, 1973.

"Quantitative model for the effects of stimulus frequency upon synchronization of auditory nerve discharges," by D. J. Anderson. J ACOUST SOC AM 54:361-364, August, 1973.

"Radiation therapy of tumors of the eighth nerve sheath," (proceedings), by H. Newman, et al. AM J ROENTGENOL RADIUM THER NUCL MED 120:562-567, March, 1974.

"Recovery from sound exposure in auditory-nerve fibers," by E. Young, et al. J ACOUST SOC AM 54:1535-1543, December, 1973.

"Recovery of detection probability following sound exposure: comparison of physiology and psychophysics," by E. Young, et al. J ACOUST SOC AM 54:1544-1553, December, 1973.

"Specific acute losses of vestibular function in four patients following unilateral section of one or all components of the eighth cranial nerve," by E. F. Miller II, et al. ANN OTOL RHINOL LARYNGOL 83:525-536, July-August, 1974.

"State of the acoustic analyzer in stuttering," by A. G. Rakhmilevich. ZH USHN NOS GORL BOLEZN 33:27-31, September-October, 1973.

"Studies on the diagnostic significance of the galvanic test in acoustic neurinoma," by C. R. Pfaltz. ARCH KLIN EXP OHREN NASEN KEHLKOPFHEILKD 205:130-134, December 17, 1973.

"The subtemporal transtentorial approach for large acoustic nerve tumors," by F. Garcia-Bengochea, et al. ACTA NEUROL LAT AM 18: 344-354, October-December, 1972.

"Symptomatology, therapy and prognosis of acoustic neurinoma (a critical report on 118) cases," by U. Borchardt, et al. PSYCHIATR NEUROL MED PSYCHOL 25:472-487, August, 1973.

"The symptoms of neuro-fibroma of the 8th nerve," by C. Ozsahinoglu, et al. J LARYNGOL OTOL 88:493-502, June, 1974.

"Synaptic transmission from hair cells to auditory fibers," (proceedings), by T. Furukawa, et al. J PHYSIOL SOC JAP 35:534, August-September, 1973.

"Tails of tuning curves of auditory-nerve fibers," by N. Y. Kiang, et al. J ACOUST SOC AM 55:620-630, March, 1974.

"Technique and value of gas and Pantopaque cisternography in the diagnosis of cerebello-pontine angle tumours," by S. Wende, et al. NEURORADIOLOGY 2:24-29, March, 1971.

"Theory of binaural interaction based on auditory-nerve data. I. General strategy and preliminary results on interaural discrimination," by H. S. Colbrun. J ACOUST SOC AM 54:1458-1470, December, 1973.

"Ulcero-mutilating acropathies and diabetic perforating plantar diseases,"

ACOUSTIC NERVE

(letter), by C. Schmidt, et al. NOUV PRESSE MED 3:528, March 2, 1974.

"Ultrasound perception in pathologic processes of the cerebello-pontine angle area," by N. S. Blagoveshchenskaia, et al. VOPR NEIROKHIR 37:18-22, September-October, 1973.

"Undiagnosed acoustic neurinomas. A presentation of 4 cases," by J. Thomsen, et al. ARCH KLIN EXP OHREN NASEN KEHLKOPF-HEILKD 204:175-182, October 12, 1973.

"Unexpected discovery of 2 neurinomas of the 8th pair in transtemporal surgery," by R. De Asis Alonso, et al. ACTA OTORINOLARYN-GOL IBER AM 23:840-848, 1972.

"An unusual acoustic neurinoma localized between brain stem and basilar artery using emulsified pantopaque cisternograpny," by J. L. Fox, et al. SURG NEUROL 2:329-332, September, 1974.

"Use of Xavin in auditory nerve lesions," by S. Spellenberg. THER HUNG 19:53-58, 1971.

"Value of foot dearterialization in ulceromutilating acropathies," (letter), by C. Lefaucher, et al. NOUV PRESSE MED 2:2958, December 8, 1973.

"Vestibular disturbances in the early diagnosis of tumors of the acoustic nerve," by L. B. Jongkees. J FR OTORHINOLARYNGOL 22:787-794, November, 1973.

ACOUSTIC STIMULATION
see also: Noise Research

"Acoustic conduction of the auditory ossicles," by E. I. Volpliushkin, et al. ZH USHN NOS GORL BOLEZN 33:12-15, 1973.

"Acoustic jaw reflex in man: its relationship to other brain-stem and microreflexes," by K. Meier-Ewert, et al. ELECTROENCEPHA-LOGR CLIN NEUROPHYSIOL 36:629-637, June, 1974.

"Acoustically evoked potentials (a.e.p.) in neonates with special consideration in intrauterine dystrophis," by G. Muller, et al. KLIN PAEDIATR 185:449-457, November, 1973.

"Advantages of quasi-simultaneous stimulation in ERA," by D. Krell, et al. AUDIOLOGY 13(4):342-348, 1974.

"Analysis of acoustic signal registered during respirofonometry," by I. Simacek. CAS LEK CESK 113(37):1122-1124, September 13, 1974.

"Analysis of central nervous system involvement in the microwave auditory effect," by E. M. Taylor, et al. BRAIN RES 74:201-208, July 12, 1974.

"An analysis of sensory interaction," by R. L. Taylor. NEUROPSYCHOLOGIA 12:65-71, January, 1974.

"Apparatus for studying the hearing function by means of impedance measuring," by B. S. Moroz, et al. ZH USHN NOS GORL BOLEZN 0(1):115-117, January-February, 1974.

"Articulatory interpretation of the 'singing formant'," by J. Sundberg. J ACOUST SOC AM 55:838-844, April, 1974.

"Attempt at physical characterization of the passive sound behavior in the lung on a model," by H. R. Bohme. Z GESAMTE INN MED 29:401-406, May 15, 1974.

"The auditory stimuli to evoke a clear average response at behavioral threshold," by S. Funasaka, et al. AUDIOLOGY 13:162-172, March-April, 1974.

"Averaged electroencephalographic response to intensity modulated tone," by T. Okitsu. TOHOKU J EXP MED 112:315-323, April, 1974.

"Bayesian density functions for Gaussian pulse shapes in Gaussian noise," by R. Clow, et al. IEEE PROC 62:134-136, January, 1974.

"Calculation of an equivalent level of nonstable noise," by E. I. Denisov, et al. GIG TR PROF ZABOL 17:50-51, July, 1973.

"Changes of hippocampal single-cell activity in emotion and motivation-active stimuli," by U. Zippel, et al. ACTA BIOL MED GER 31:841-851, 1973.

"Combined action of ultrasonics and noise of standard parameters," by A. V. Il'nitskaia, et al. GIG SANIT 38:50-53, May, 1973.

"Computer program sequence for collection, reduction, analysis, and summary of auditory evoked potential data," by M. I. Mendel, et al. COMPUT BIOMED RES 6:578-587, December, 1973.

"Computerized classification of the results of screening audiometry in groups of persons exposed to noise," by I. Klockhoff, et al. AUDIOLOGY 13(4):326-334, 1974.

"Correlation between sound pressure and intra-thoracic pressure at onset of phonation," by F. Klingholz, et al. FOLIA PHONIATR 24:381-386, 1972.

"Cortical lesions: flavor illness and noise-shock conditioning," by W. G. Hankins, et al. BEHAV BIOL 10:173-181, February, 1974.

"Development of some mechanisms useful in sound localization," by S. D. Erulkar. FED PROC 33:1928-1932, August, 1974.

"Dichotic competition of simultaneous tone bursts of different frequency. I. Dissociation of pitch from lateralization and loudness," by R. Efron, et al. NEUROPSYCHOLOGIA 12:249-256, March, 1974.

"A 'distraction effect' of noise bursts," by S. Fisher. PERCEPTION 1:223-236, 1972.

"Distribution of the changes in the receptor auditory cells along the basilar membrane of the cochlea under the influence of narrow-band (octave) noise of different frequency characteristics," by V. F. Anichin, et al. ZH USHN NOS GORL GOLEZN 33:15-20, 1973.

"Does tonotopicity subserve the perceived elevation of a sound? by R. A. Butler. FED PROC 33:1920-1923, August, 1974.

"Early averaged electroencephalic responses to clicks in neonates," by C. C. McRandle, et al. ANN OTOL RHINOL LARYNGOL 83(5): 695-702, September-October, 1974.

"Effect of contralateral broad-band noise on frequency discrimination," by J. M. Labiak, et al. ACTA OTOLARYNGOL 77:29-36, January-February, 1974.

"The effect of noise on visual fields," by J. E. Letourneau, et al. EYE EAR NOSE THROAT MON 53:49-51, February, 1974.

"Effects of acoustical stimulation on equilibrium," by S. L. Vanderhei. J ACOUST SOC AM 55:Suppl:41, 1974.

"Effects of frequency modulation on auditory averaged evoked response," by M. L. Lenhardt. AUDIOLOGY 10:18-22, January-February, 1971.

"Effects of low—pass filtering on the rate of learning and retrieval from memory of speech-like stimuli," by R. Novak, et al. J SPEECH HEAR RES 17:279-285, June, 1974.

"Effects of masking-spectrum slope and interaural phase on detection of tones," by M. Sonn. PERCEPT MOT SKILLS 38:776-784, June, 1974.

"Electrocochleography in clinical-audiological diagnosis," by H. Sohmer, et al. ARCH OTORHINOLARYNGOL 206:91-102, March 25, 1974.

"Electrographic correlates of lateral asymmetry in the processing of verbal and nonverbal auditory stimuli," by H. Neville. J PSYCHOLINGUIST RES 3:151-163, April, 1974.

"Equal aversion levels for pure tones and one third-octave bands of noise," by J. A. Molino. J ACOUST SOC AM 55:1285-1289, June, 1974.

"Equating individual differences for auditory input," by D. McGuinness. PSYCHOPHYSIOLOGY 11:113-120, March, 1974.

"Evoked response thresholds for long and short duration tones," by C. T. Grimes, et al. AUDIOLOGY 10:358-364, September-December, 1971.

"Extraversion and auditory sensitivity to high and low frequency," by R. M. Stelmack, et al. PERCEPT MOT SKILLS 38:875-879, June, 1974.

"Functional state of the hearing analyzer in the concurrent action of noise, vibration, physical work and high temperature," by M. V. Ratner, et al. GIG TR PROF ZABOL 16:47-49, July, 1972.

"Generality of interference by tonal stimuli in recognition memory for pitch," by D. Deutsch. Q J EXP PHYSIOL 26:229-234, May, 1974.

"Graphic method of determining the distance from a noise source to the area the level is standardized," by A. L. Vasil'eva, et al. GIG SANIT 38:84-86, March, 1973.

"Homolateral and contralateral masking of tinnitus by noise-bands and by pure tones," by H. Feldmann. AUDIOLOGY 10:138-144, May-June, 1971.

"Hypertensive effects of prolonged auditory, visual, and motion stimulation," by H. H. Smookler, et al. FED PROC 32:2105-2110, November, 1973.

"Identification of renal calculi by a new sonar blunt curette," by H. Tammen. UROL INT 28:158-160, 1973.

"Increased adult auditory responsiveness resulting from juvenile acoustic experience," by K. R. Henry. FED PROC 32:2098-2100, November, 1973.

"Infrasound." (editorial). LANCET 2:1368-1369, December 15, 1973.

"Infrasound—occurrence and effects," by S. Handel, et al. LAKARTID-

NINGEN 71:1635-1639, April 17, 1974.

"Interactions and range effects in experiments on pairs of stresses: mild heat and low-frequency noise," by E. C. Poulton, et al. J EXP PSYCHOL 102:621-628, April, 1974.

"Interaural alternation and speech intelligibility," by C. Speaks, et al. J ACOUST SOC AM 56(2):640-644, August, 1974.

"Intracellular electric responses to sound in a vertebrate cochlea," by M. J. Mulroy, et al. NATURE 249:482-485, May 31, 1974.

"Intersensory facilitation, errors and corrections in a discontinuous tracking task," by A. Semjen. ANNEE PSYCHOL 73:403-417, 1973.

"Is frequency information extracted from electrical stimulation of the auditory system?" by F. W. Mis, et al. EXP NEUROL 43:227-241, April, 1974.

"Laboratory note. Scalp-recorded early responses in man to frequencies in the speech range," by G. Moushegian, et al. ELECTROENCEPHALOGR CLIN NEUROPHYSIOL 35:665-667, December, 1973.

"Loss and recovery processes operative at the level of the cochlear microphonic during intermittent stimulation," by G. R. Price. J ACOUST SOC AM 56:183-189, July, 1974.

"Loudness discomfort level: selected methods and stimuli," by D. E. Morgan, et al. J ACOUST SOC AM 56(2):577-581, August, 1974.

"Loudness discomfort level under earphone and in the free field: the effects of calibration methods," by D. E. Morgan, et al. J ACOUST SOC AM 56:172-178, July, 1974.

"Loudness of brief tones in hearing-impaired ears. Temporal integration of acoustic energy at suprathreshold levels in patients with presbyacusis," by C. B. Pedersen, et al. ACTA OTOLARYNGOL 76:402-409, December, 1973.

"Loudness tracking and the staircase method in the measurement of adaptation," by T. E. Stokinger, et al. AUDIOLOGY 11:161-168, May-August, 1972.

"Masking produced by sinusoids of slowly changing frequency," by D. A. Ronken. J ACOUST SOC AM 54:905-915, October, 1973.

"Method of improving the noise stability of a magnetic recording system in registering biomedical information," by O. V. Balabanov, et al. VESTN AKAD MED NAUK SSSR 28:61-64, 1973.

"Microwave hearing: evidence for thermoacoustic auditory stimulation by pulsed microwaves," by K. R. Foster, et al. SCIENCE 185: 256-258, July 19, 1974.

"Model for wave propagation in a lossy vocal tract," by M. M. Sondhi. J ACOUST SOC AM 55:1070-1075, May, 1974.

"Neurophysiological mechanisms of sound localization," by A. Starr. FED PROC 33:11911-1914, August, 1974.

"Noise coupling between accommodation and accommodative vergence," by D. Wilson. VISION RES 13:2505-2513, December, 1973.

"Nonauditory of effects of noise," by M. Schiff. TRANS AM ACAD OPHTHALMOL OTOLARYNGOL 77:1384-1398, September-October, 1973.

"A non-verbal analogue to the verbal transformation effect," by N. J. Lass, et al. CAN J PSYCHOL 27:272-279, September, 1973.

"Platelet adhesiveness during noise exposure," by B. Maass, et al. DTSCH MED WOCHENSCHR 98:2153-2155, November 9, 1973.

"Pitch and stimulus fine structure," by F. L. Wightman. J ACOUST SOC AM 54:397-406, August, 1973.

"Polysensory interactions in the cuneate nucleus," by S. F. Atweh, et al. J PHYSIOL 238:343-355, April, 1974.

"A possible neurophysiological basis for the precedence effect," by I. C. Whitfield. FED PROC 33:1915-1916, August, 1974.

"The recovery cycle of the averaged auditory evoked response during sleep in normal children," by E. M Ornitz, et al. ELECTROENCEPHALOGR CLIN NEUROPHYSIOL 37:113-122, August, 1974.

"Relations between the psychophysics and the neurophysiology of sound localization," by G. Moushegian, et al. FED PROC 33:1924-1927, August, 1974.

"Role of centrifugal pathways to cochlear nucleus in detection of signals in noise," by J. O. Pickles, et al. J NEUROPHYSIOL 36:1131-1137, November, 1973.

"Role of the fornix in the septal syndrome," by D. S. Olton, et al. PHYSIOL BEHAV 13(2):269-279, August, 1974.

"Slow cortical evoked potentials: interactions of auditory, vibro-tactile and shock stimuli," by I. S. Hay, et al. AUDIOLOGY 10:9-17, January-February, 1971.

"Some observations on the effects of attention to stimuli on the amplitude of the acoustically evoked response," by L. W. Keating, et al. AUDIOLOGY 10:177-184, May-June, 1971.

"Sound detector of X-ray and gamma-radiation," by T. V. Lapchuk, et al. MED RADIOL 19(7):77-82, July, 1974.

"Stimulation interval-dependent acoustic evoked potentials," by E. Sturzebecher, et al. ACTA OTOLARYNGOL 77:256-260, April, 1974.

"A study of laser-acoustic air pollution monitors," by L. G. Rosengren, et al. J PHYS E SCI INSTRUM 7:125-133, February, 1974.

"Synthesis of low-frequency noise for use in biological experiments," by A. S. French. IEEE TRANS BIOMED ENG 21:251-252, May, 1974.

"A technique for individual noise exposure assessment," by R. Colman, et al. J ACOUST SOC AM 55(2):464, 1974.

"TLVs threshold limit values for physical agents adopted by ACGIH for 1973." J OCCUP MED 16:49-58, January, 1974.

"TMJ sound prints: electronic auscultation and sonagraphic audiospectral analysis of the temporomandibular joint," by P. L. Ouellette. J AM DENT ASSOC 89(3):623-628, September, 1974.

"Transformation function of the external ear in response to impulsive stimulation," by G. R. Price. J ACOUST SOC AM 56:190-194, July, 1974.

"The use of alternated stimuli to reduce response decrement in the auditory testing of newborn infants." by D. Ling, et al. J SPEECH HEAR RES 14:531-534, September, 1971.

"Voltage noise, current noise and impedance in space clamped squid giant axon," by E. Wanke, et al. PFLUEGERS ARCH 347:63-74, February 18, 1974.

ACOUSTICAL SOCIETY OF AMERICA
"Meeting, 86th, Los Angeles, Oct. 30-Nov. 2; program and abstracts of papers." ACOUSTICAL SOC AM J 55:383-492, February, 1974.

"MM11 at the 86th meeting of ASA," by S. R. Lane. J ACOUST SOC AM 55:1346-1348, June, 1974.

"Rebuttal to papers on aircraft noise by S. R. Lane at the 86th Meeting of the ASA," by V. E. Callaway. J ACOUST SOC AM 55:1343-1345, June, 1974.

"Reply to criticisms by V. E. Callaway of papers MM1 and MM11 at the 86th meeting of the ASA," by S. R. Lane. J ACOUST SOC AM 55:1346-1348, June, 1974.

AIR CONDITIONING
"Acoustics in air conditioning." HOSP ENG 27:8-19, January, 1973.

AIR CONDITIONING

"Better method for air system acoustical design," by J. A. Reese, et al. ASHRAE J 16:59-63, September, 1974.

"Compressors beat new noise-law levels." ELEC WORLD 180:85, July 15, 1973.

"Consultants says duct noise regeneration bothers owners, needs A/C industry study," by M. Kodaras. AIR COND HEAT & REFRIG N 130:13, October 1, 1973.

"Contractors: protection clause makes owner pay to quell noise complaints." AIR COND HEAT & REFRIG N 131:29, February 25, 1974.

"For noise control, absorption method called least effective." AIR COND HEAT & REFRIG N 133:20, December 9, 1974.

"How Day & Night-Payne reduced sound, increased efficiencies." AIR COND HEAT & REFRIG N 131:27, February 18, 1974.

"Milwaukee adopts noise law for home air conditioners." AIR COND HEAT & REFRIG N 130:1 plus, September 3, 1973.

"Pinellas County's noise ordinance sets A/C sound levels at 60 dBA," by P. Kelsey. AIR COND HEAT & REFRIG N 133:3, November 11, 1974.

"Sound-power testing experiences of an independent laboratory," by M. J. Kodaras, et al. ACOUSTICAL SOC AM J 54:956-959, October, 1973.

"Unitary mfrs detail price, time penalties of quieter equipment." AIR COND HEAT & REFRIG N 130:27, December 3, 1973.

AIRCRAFT NOISE

"Aircraft environmental problems," by V. L. Blumenthal, et al. J AIRCRAFT 10:529-537, September, 1973.

"Aircraft noise abatement via annex 16 of the Chicago convention—a viable alternative," by S. S. Kalsi. TEX INT L J 9:1-18, Winter, 1974.

"Aircraft noise and psychiatric morbidity," by F. Gattoni, et al. PSYCHOL MED 3:516-520, November, 1973.

"Aircraft noise induced vibration in fifteen residences near Seattle-Tacoma International Airport," by S. M. Cant, et al. AM IND HYG ASSOC J 34:463-468, October, 1973.

"Aircraft noise now on-line to Stockholm's new pollution monitoring network." ATMOSPHERIC ENVIRONMENT 7:Suppl:ii-iii, December, 1973.

"Airport officials hit FAA, DOT on noise, fund issues," by E. J. Bulban. AVIATION W 99:34 plus, October 29, 1973.

"Annoyance judgments of aircraft with and without acoustically treated nacelles," by P. N. Borsky. J ACOUST SOC AM 55:Suppl:67, 1974.

"California court bars class action; suit to recover damages for aircraft noise." AVIATION W 101:35, October 7, 1974.

"Certification of Concorde, Tu-144 backed by environmental unit." AVIATION W 101:297, July 15, 1974.

"City of Burbank v. Lockheed Air Terminal, Inc. (93 Sup Ct 1854): federal preemption of aircraft noise regulations and the future of proprietary restrictions." NYU REV L & SOC CHANGE 4:99-113, Winter, 1974.

"Community response to elimination of nighttime aircraft noise," by H. G. Smith. J ACOUST SOC AM 55:Suppl:68, 1974.

"Constitutional law: aircraft noise control preempted by federal government." WASHBURN L J 13:118-123, Winter, 1974.

"A cost-effective method of evaluating aircraft noise-abatement options [San Antonio International Airport]," by S. R. Goldberg. TEX BUS R 47:284-287, December, 1973.

"Crash resistance and noise improvements in business aircraft." AUTOMOTIVE ENG 82:57-59, April, 1974.

"Effects of airplane noise on health: an examination of three hypotheses," by D. B. Graeven. J HEALTH & SOC BEHAV 15:336-343, December, 1974.

"Effects of a traffic noise background on judgements of aircraft noise," by C. A. Powell. J ACOUST SOC AM 55:Suppl:68, 1974.

"Engine-over-the-wing noise research," by M. Reshotko, et al. J AIRCRAFT 11:195-196, April, 1974.

"Environmental law—aircraft noise control—use of local police powers to impose curfews on air flights is preempted by the federal aviation act of 1958 as amended by the noise control act of 1972." RUTGERS CAMDEN L J 5:566-584, Spring, 1974.

"Environmental law—aircraft noise regulation—federal pre-emption." NY L F 20:165-176, Summer, 1974.

"Environmental law—federal regulation of aircraft noise under federal aviation act precludes local police power noise restrictions." BC IND & COM L R 15:848-862, April, 1974.

"Environmental law—federal regulation under federal aviation act and noise control act preempts the field of airport and aircraft noise control rendering local airport curfews invalid." KAN L REV 22:319-336, Winter, 1974.

"EPA readies noise recommendations," by R. K. Ellingsworth. AVIATION W 100:32-33, March 25, 1974.

"FAA gets contradictory noise guidance," by W. A. Shumann. AVIATION W 101:40-42, August 12, 1974.

"FAA noise proposal draws big response from citizen groups." AVIATION W 101:26, July 22, 1974.

"FAA proposes to quiet jets." ASTRONAUTICS & AERONAUTICS 12:13, May, 1974.

"FAA pushing nacelle noise retrofits," by W. A. Shumann. AVIATION

W 100:29-30, May 27, 1974.

"Federal pre-emption—aviation noise control—the Federal aviation administration, monitored by the environmental protection agency, has full control over aviation noise, pre-empting state and local control, including a municipal ordinance which imposed a curfew on certain jet take-offs during certain night-time hours." J AIR L 40:341-349, Spring, 1974.

"F28; development of the MK 5000/6000; further development of the MK 1000." AIRCRAFT ENG 45:14-19, October, 1973.

"House unit urges FAA to delay rule on quiet nacelle retrofit." AVIATION W 100:24, June 24, 1974.

"H-P noise monitor fails, repay $178K to LA airport," by J. Fraser. ELECTRONIC N 19:1 plus, March 25, 1974.

"Human perception of apparent direction and movement of aircraft noise," by W. J. Gunn, et al. J ACOUST SOC AM 55:Suppl:68, 1974.

"Insulating against jet noise." CHEMISTRY 47:21, February, 1974.

"Jet noise at schools near Los Angeles International Airport," by S. R. Lane, et al. J ACOUST SOC AM 56:127-131, July, 1974.

"Key noise reduction decisions imminent," by W. A. Shumann. AVIATION W 100:31-33, January 7, 1974.

"Measured variations in aircraft noise near Arlanda airport," by A. R. Kajland. ACOUSTICAL SOC AM J 56:329-331, August, 1974.

"MM11 at the 86th meeting of the ASA," by S. R. Lane. J ACOUST SOC AM 55:1346-1348, June, 1974.

"NASA JT8D refan program nears end," by M. L. Yaffee. AVIATION W 101:46-47, July 22, 1974.

"Noise; future targets," by G. M. Lilley. AERONAUTICAL J 78:459-

463, October, 1974.

"Pilatus develops quiet turbo Porter." AVIATION W 99:65-67, October 29, 1973.

"Property—eminent domain—even where no direct overflight occurs, aircraft noise and air pollution depriving property owners of the practical use and enjoyment of their land is a taking requiring compensation for the diminution of the land's market value." J URBAN L 52:636-648, Winter, 1974.

"Propulsion system design for the ATT," by G. L. Brines. J AIRCRAFT 10:487-490, August, 1973.

"Putting all our noise technology to work," by R. P. Jackson. ASTRONAUTICS & AERONAUTICS 12:48-51, January, 1974; Discussion 12:4-6, April, 1974.

"Rebuttal to papers on aircraft noise by S. R. Lane at the 86th Meeting of the ASA," by V. E. Callaway. J ACOUST SOC AM 55:1343-1345, June, 1974.

"Reduction of VTOL operational noise through flight trajectory management," by F. H. Schmitz, et al. J AIRCRAFT 10:385-394, July, 1973.

"Reply to criticisms by V. E. Callaway of papers MM1 and MM11 at the 86th meeting of the ASA," by S. R. Lane. J ACOUST SOC AM 55:1346-1348, June, 1974.

"Silencer for jet whine." NEWSWEEK 84-34, July 1, 1974.

"The social impact of aircraft noise," by A. Alexandre. TRAFFIC Q 28:371-388, July, 1974.

"SST noise levels called acceptable." AVIATION W 100:26, March 18, 1974.

"Stronger rules recommended to cut propeller aircraft noise." AVIATION W 100:61, March 25, 1974.

"Takahashi's work on jet noise," by S. Kondo, et al. J HUM ERGOL 2:93-94, September, 1973.

"Thrust reverser noise estimation," by M. R. Fink. J AIRCRAFT 10: 507-508, August, 1973.

"Unified approach to aerodynamic sound generation in the presence of solid boundaries," by M. Goldstein. ACOUSTICAL SOC AM J 56: 497-500, August, 1974.

"U.S. officials find Concorde noise acceptable." AVIATION W 100: 27, June 24, 1974.

"VFW 614; the quietest jet transport in the world." AIRCRAFT ENG 46:5, June, 1974.

AIRPLANES

"Business flying faces new challenges; Hawker Siddeley, rolls to test HS 125 with sound suppression," by H. J. Coleman. AVIATION W 99:57-58, September 24, 1973.

"The hazardous noise exposure to which airline passengers are subjected," by S. R. Lane. J ACOUST SOC AM 55(2):465, 1974.

"Judged acceptability of noise exposure during television viewing," by L. E. Langdon, et al. ACOUSTICAL SOC AM J 56:510-515, August, 1974.

AIRPLANES: MODEL

AIRPORTS

"Aircraft noise induced vibration in fifteen residences near Seattle-Tacoma International Airport," by S. M. Cant, et al. AM IND HYG ASSOC J 34:463-468, October, 1973.

"Airport noise unaffected by flight cuts," by W. A. Shumann. AVIATION W 100:29-30, February 4, 1974.

"Airport officials hit FAA, DOT on noise, fund issues," by E. J. Bulban. AVIATION W 99:34 plus, October 29, 1973.

"California court bars class action; suit to recover damages for aircraft noise." AVIATION W 101:35, October 7, 1974.

"A cost-effective method of evaluating aircraft noise-abatement options [San Antonio International Airport]," by S. R. Goldberg. TEX BUS R 47:284-287, December, 1973.

"EPA readies noise recommendations," by R. K. Ellingsworth. AVIATION W 100:32-33, March 25, 1974.

"Federal pre-emption and airport noise control." URBAN L ANN 8: 229-239, 1974.

"Federal preemption in airport noise abatement regulation: allocation of federal and state power." MAINE L REV 26:321-344, 1974.

"H-P noise monitor fails, repay $178K to LA airport," by J. Fraser. ELECTRONIC N 19:1 plus, March 25, 1974.

"Insulating against jet noise." CHEMISTRY 47:21, February, 1974.

"Inverse condemnation and nuisance: alternative remedies for airport noise damage." SYRACUSE L REV 24:793-809, 1973.

"Jet noise at schools near Los Angeles international airport," by S. R. Lane, et al. ACOUSTICAL SOC AM J 56:127-131, July, 1974.

"Massport vs. community; Logan international airport expansion controversy," by D. Nelkin. SOCIETY 11:27-36 plus, May, 1974.

"Measured variations in aircraft noise near Arlanda airport," by A. R. Kajland. ACOUSTICAL SOC AM J 56:329-331, August, 1974.

"No place to land: the airport crisis." READ DIGEST 102:101-105, March, 1973.

"On the hearing of residents near airports," by D. C. Nagel, et al. J ACOUST SOC AM 55(2):463-464, 1974.

"Planning for airports in urban environments—a survey of the problem

AIRPORTS

and its possible solutions," by M. L. Dworkin. TRANSP L J 5: 183-214, July, 1973.

AMPLITUDE DISCRIMINATION

ANGIOTENSIN

ARCHITECTURE

"ABC's of sound reinforcement," by M. Koller. RADIO-ELECTRONICS 45:40, August, 1974.

"Acoustics of educational facilities," by E. P. Caffarella, Jr. AV INSTR 18:10-11, December, 1973.

"Directory of graduate education in acoustics," by W. M. Wright, et al. J ACOUST SOC AM 55:1105-1115, May, 1974.

"New multipurpose Hall; theater and broadcasting facilities," by H. Moriyama. SMPTE J 83:169-175, March, 1974.

"Speakers and rooms," by R. Hodges. POP ELECTR 6:22-26, August, 1974.

"Speech acoustics for the theatre," by D. L. Klepper. AUDIO ENG SOC J 22:15-19, January-February, 1974.

"Unobtrusive sound reinforcement for an open-plan school." ARCHIT REC 156:151-152, September, 1974.

"What's so important about acoustics?" by A. Tipton. SCH MUS 45: 18 plus, May, 1974.

AUTOMOBILE NOISE
see also: Traffic Noise

"Conference on Vehicle noise and the designer, Hatfield, England." ENGINEER 238:27, May 2, 1974.

"Europe seeks common solutions to problems of emissions and noise."

AUTOMOBILE NOISE

AUTOMOTIVE ENG 81:25-31, March, 1973.

"The measurement of noise from moving vehicles," by E. L. Hixson, et al. J ACOUST SOC AM 54(1):332, 1973.

"Motorway noise and dwellings." BUILD RES ESTAB DIGEST 153:1-7, May, 1973.

"Noise abatement, problems and progress, symposium." DIESEL EQUIP SUPT 52:42-44, June, 1974.

"Vehicular noise regulation in Hawaii," by J. C. Burgess. ACOUSTICAL SOC AM J 56:905-910, September, 1974.

BAROTRAUMA

BEHAVIOR
see also: Learning
Psychology

"Application of some new survey techniques for assessing exposure to noise and human reaction," by M. Braden, et al. J ACOUST SOC AM 55(2):464, 1974.

"Effect of ambient illumination on noise level of groups," by M. Sanders, et al. J APP PSYCHOL 59:527-528, August, 1974.

"Effects of noise on people," by J. D. Miller. ACOUSTICAL SOC AM J 56:729-764, September, 1974.

"A field investigation by hypnosis of sound loci importance in human behavior," by M. H. Erickson. AM J CLIN HYPN 16:92-100, October, 1973.

"The 'soft-spoken' woman. II. Auditory vs non-auditory monitoring of loudness behaviors," by P. J. Dembowski, et al. J COMMUN DISORD 6:206-212, September, 1973.

BEKESY TYPING
see: Inner Ear

BOILERS

BREATH
"Breath-sound changes after cigarette smoking," (letter), by C. W. Laird, et al. LANCET 1:808, April 27, 1974.

BUS NOISE
"Shielding cuts truck and bus noise." AUTOMOTIVE ENG 81:15-16, August, 1973.

CAFFEINE CITRATE

CARDIOVASCULAR SYSTEM
see also: Noise Research

"Change in heart rate due to acoustic stimulation, audiologic test-method," (proceedings), by H. Chuden. ARCH KLIN EXP OHREN NASEN KEHLKOPFHEILKD 205:231-238, December 17, 1973.

"Changes of heart rate associated with responses to cyclic visual and acoustic stimuli," by K. Scheuch, et al. ACTA BIOL MED GER 32(4):385-391, 1974.

"Characteristics of the cardiovascular system of adolescent workers subjected to the action of stable industrial noise," by E. A. Gel'tishcheva. GIG TR PROF ZABOL 16:29-33, July, 1973.

"Complex effects of different factors (noise, tranquilizer, difficult working condition, test time) on pursuit tracking performance and beat-to-beat heart rate behavior," by H. Strasser, et al. INT ARCH ARBEITSMED 31:81-103, May 23, 1973.

"Modifications of epinephrine, norepinephrine, blood lipid fractions and the cardiovascular system produced by noise in an industrial medium," by G. A. Ortiz, et al. HORM RES 5:57-64, January, 1974.

"Sound spectrographic examinations of heart sounds and murmurs in aortic valve lesions," by A. Aigner, et al. Z KARDIOL 63(3):269-278, March, 1974.

CATARACTS

CHILDREN
"Increased adult auditory responsiveness resulting from juvenile acoustic experience," by K. R. Henry. FED PROC 32:2098-2100, November, 1973.

"The recovery cycle of the averaged auditory evoked response during sleep in normal children," by E. M. Ornitz, et al. ELECTROEN-CEPHALOGR CLIN NEUROPHYSIOL 37:113-122, August, 1974.

"School children in London; noise survey." HEALTH VISIT 45:274-277, September, 1972.

"Some implications regarding high frequency hearing loss in school-age children," by R. L. Cozad, et al. J SCH HEALTH 44:92-96, February, 1974.

CHURCHES

CIRCULATORY SYSTEM
see: Noise Research

CLINDAMYCIN

CLINICAL ASPECTS
"Acoustic neurinoma. A comparison of the clinical picture and the electroencephalogram," by Z. Mensikova, et al. SB VED PR LEK FAK KARLOVY UNIV 15:401-408, 1972.

"Comprehensive clinical and psychological studies of patients exposed to chronic acoustic trauma," by S. Klonowski, et al. POL TYG LEK 29:313-315, February 25, 1974.

"Electrocochleography in clinical-audiological diagnosis," by H. Sohmer, et al. ARCH OTORHINOLARYNGOL 206:91-102, March 25, 1974.

"Preliminary summary on the clinical experience with treatment of 77 cases of explosion deafness." CHIN MED J 4:238-241, 1974.

COMBUSTION NOISE
"Combustion noise," by F. E. J. Briffa, et al. COMBUSTION 45:27-37, March, 1974.

COMMUNITY NOISE
see also: Urban Noise

"Community noise control [South Florida]," by S. E. Dunn. FLA ENVIRONMENTAL & URBAN ISSUES 1:7-10, October-November, 1973.

"Community noise survey of Medford, Massachusetts," by J. E. Wesler. J ACOUST SOC AM 54:985-995, October, 1973.

"Monitoring community noise," by M. C. Branch, et al. AM INST PLAN J 40:266-273, July, 1974.

"Possibilities and problems of achieving community noise acceptance of VTOL," by W. Z. Stepniewski, et al. AERONAUTICAL J 77:311-326, June, 1973.

"Some techniques for assessment of community noise environments," by L. C. Sutherland, et al. J ACOUST SOC AM 55(2):464, 1974.

"Sound-level measurements in the community," by D. S. Allen. AIR COND HEAT & REFRIG N 131:30-32, March 4, 1974.

COMPRESSED AIR NOISE
"Compressor sound control," by G. M. Diehl. TAPPI 57:75, April, 1974.

"Muffling techniques for reducing pneumatic tool noise," by R. A. Willoughby, et al. PLANT ENG 27:109-111, September 6, 1973.

"Noise and pile driving," by D. J. Hagerty, et al. ROADS & STS 117:70-71, August, 1974.

"Porous plastic silencers hush pneumatic presses." ENGINEER 237:23, October 18, 1973.

COMPRESSED AIR NOISE

"Quieted air cylinders meet OSHA noise requirements." HYDRAULICS & PNEUMATICS 26:133, September, 1973.

"Silent air; Atlas Copco Silensair STS 71." ENGINEERING 213:425, June, 1973.

COMPRESSOR NOISE
see: Compressed Air Noise

DENTAL NOISE
"Noise from turbine drill. Risk for hearing injuries among dentists?" by T. Hundseth, et al. NOR TRANNLAEGEFOREN TID 83:185-187, May, 1973.

"Sources of noise and lowering of the noise level in the dental office," by W. Hoefig. DTSCH ZAHNAERZTL Z 99:172-175, February, 1973.

DIABETES

DIAZEPAM

DISCOTHEQUES

DRUGS
see: Under Specific Drug

EARMUFFS
"Ear muffs made for all-day wear." PURCHASING 77:71, December 3, 1974.

EARPHONES

EARPLUGS
"Antinoise ear plug made of film porolon," by V. Ia. Gapanovich, et al. GIG TR PROF ZABOL 16:54, July, 1972.

"Determination of in-use attenuation value for selected ear plugs," by C. E. Scott III, et al. J ACOUST SOC AM 54(1):328, 1973.

ECONOMIC ASPECTS

"Big noises are being heard as industry considers the cost," by C. Beatson. ENGINEER 238:38-39, February 7, 1974.

"A cost-effective method of evaluating aricraft noise-abatement options [San Antonio international airport]," by S. R. Goldberg. TEX BUS R 47:284-287, December, 1973.

"Expensive sound of silence." BUS W 28, July 20, 1974.

"Fragmented noise control sought; EPA to press for noise funding scheme," by C. E. Schneider. AVIATION W 99:19-20, July 9; 32-33, July 16, 1973.

"Low-cost approach to area-wide noise monitoring," by T. E. Siddon, et al. J ACOUST SOC AM 54:646-649, September, 1973.

"Should the practice for financial compensation for occupational acoustic trauma be changed?" by A. Ahlmark, et al. LAKARTIDNINGEN 70:3151-3154, September 12, 1973.

"Widened practice for compensation justified—temporary solutions must be possible to create," by A. Ahlmark, et al. LAKARTIDNINGEN 71:823, February 27, 1974.

ELECTRONIC NOISE
see also: Music
Musical Instruments

"Electronic reverberation equipment in the Stockholm concert hall," by S. Dahlstedt. AUDIO ENG SOC J 22:627-631, October, 1974.

"Experiment on electronic noise in the freshman laboratory," by D. L. Livesey, et al. AM J PHYS 41:1364-1367, December, 1973.

"More phenomena," by C. T. T. Comber. ELECTRONIC & POWER 19:298, July 12, 1973.

"Noise limitations on measurements made with SLUGS," by J. C. Gallop. SCI INSTR 7:855-859, October, 1974.

EMPLOYEES

"Effectiveness of different ear protectors in protecting the employee from over exposure in industrial environments," by J. E. Stephenson, et al. J ACOUST SOC AM 54(1):301, 1973.

"A noise hazard to local authority employees," by W. A. Pollitt, et al. COMMUNITY HEALTH 5:19-23, July-August, 1973.

ENGINES: DIESEL

"Lowering Diesel noise through hardware modifications; fleet week preview." AUTOMOTIVE ENG 81:41-47, June, 1973.

ENGINES: JET
see: Aircraft Noise

ENVIRONMENTAL HEALTH

"Analysis of the degree of influence of environmental factors with multiple combined action," by R. E. Sova, et al. GIG TR PROF ZABOL 18:46-48, February, 1974.

"A backward glance at noise pollution," by G. Rosen. AM J PUBLIC HEALTH 64:514-517, May, 1974.

"Can we achieve a quieter environment?" by K. S. Oliphant. ASTM STAND N 2:8-13, May, 1974.

"Committee on Environmental Hazards. Noise pollution: neonatal aspects." PEDIATRICS 54(4):476-479, October, 1974.

"Control of environmental noise," by P. Jensen. J AIR POLLUT CONTROL ASSOC 23:1028-1034, December, 1973.

"Effect of environment on the health of the mining personnel in copper ore mines of the Legnica-Glogow Copper Region," by I. Juzwiak, et al. POL TYG LEK 29:811-814, May 13, 1974.

"Environment resource packets get wide use." CHEMICAL & ENGINEERING NEWS 52,4:25-26, January, 1974.

"Environmental noise," by F. Merluzzi. MED LAV 64:115-120, March-

April, 1973.

"Environmental noise level as a factor in the treatment of hospitalized schizophrenics," by M. F. Ozerengin. DIS NERV SYST 35(5):241-245, 1974.

"Hearing loss in adults: relation to age, sex, exposure to loud noise, and cigarette smoking," by A. B. Siegelaub, et al. ARCH ENVIRON HEALTH 29:107-109, August, 1974.

"How sound an environment?" by M. Cooper. DAILY TELEGRAPH 11, May 4, 1974.

"Jaundiced eye," by S. Novick. ENVIRONMENT 15:inside cover, December, 1973.

"Man and the environmental noise," by E. Wende, et al. INTERNIST 14:224-229, May, 1973.

"Noise as a pollutant," by J. Connell. DIST NURS 15:196, December, 1972.

"Noise pollution," (editorial), by P. W. Alberti. CAN J OTOLARYNGOL 1:279-280, 1972.

"Noise pollution: just how bad is it?" by J. E. Watson. MED TIMES 102: 51-59, March, 1974.

"The optimum human environment," by H. Hillman. NURS TIMES 69: 692–695, May 31, 1973.

"Sound pollution," by R. Moody. SCI & CHILD 12:6, October, 1974.

"Total environmental noise problem," by A. Glorig. ASSE J 19:22-26, July, 1974.

EPA
see: Laws and Legislation

ETHER

FAA
see: Aircraft Noise

FANS: ELECTRIC
"Calculating noise levels of fans; nomograph," by F. Caplan. PLANT ENG 28:88-89, January 24, 1974.

"Future fans: big and quiet." OIL & GAS J 72:71, June 10, 1974.

FANS: MECHANICAL
see also: Fans: Electrical

"Nomograph determines effects of fan rpm on noise level; data sheet," by F. Caplan. HEATING-PIPING 45:49-50, December 1973.

"Resilient hub hushes fan." MACHINE DESIGN 45:35, December 27, 1973.

FIREWORKS
"Noise from aerial bursts of fireworks," by D. J. Maglieri, et al. J ACOUST SOC AM 54:1224-1227, November, 1974.

FISHING VESSELS

FURNACE NOISE
see also: Combustion Noise

HEARING AIDS
"The acoustical engineer's viewpoint of hearing aid design," by S. F. Lybarger. BULL NY ACAD MED 50:917-930, September, 1974.

"Automatic sound-intensity amplifiers-storage in electric hearing aids," (proceedings), by F. J. Landwehr, et al. ARCH KLIN EXP OHREN NASEN KEHLKOPFHEILKD 205:252-256, December 17, 1973.

"Monaural and binaural speech perception through hearing aids under noise and reverberation with normal and hearing-impaired listeners," by A. K. Nabelek, et al. J SPEECH & HEARING RES 17:724-739, December, 1974.

HEARING AIDS

"Reception of consonants in a classroom as affected by monaural and binaural listening, noise, reverberation, and hearing aids," by A. K. Nabelek, et al. J ACOUST SOC AM 56(2):628-639, August, 1974.

HEARING CONSERVATION
see also: Protection

"A hearing conservation programme." HEALTH PEOPLE 8:8-9, March, 1974.

"Implementation of a hearing preservation program," by M. T. Summar. AM PAPER IND 55:16-18 plus, November, 1973.

"Program of hearing preservation in the metallurgy plant in Huachipato," by R. Benavides. REV MED CHIL 101:661-665, August, 1973.

"Role of the personnel department in a hearing conservation program," by B. Harmon. PERS J 53:531-535, July, 1974.

HEARING MEASUREMENT

HEARING STANDARDS

HELICOPTER NOISE

"Cost of noise reduction in intercity commercial helicopters," by H. B. Faulkner. J AIRCRAFT 11:89-95, February, 1974.

"Helicopter noise experiments in an urban environment," by W. A. Kinney, et al. ACOUSTICAL SOC AM J 56:332-337, August, 1974.

"Noise environment of a typical school classroom due to the operation of utility helicopters," by D. A. Hilton, et al. J ACOUST SOC AM 55:Suppl:37, 1974.

"Possibilities and problems of achieving community noise acceptance of VTOL," by W. Z. Stepniewski, et al. AERONAUTICAL J 77:311-326, June, 1973.

"Reduction of VTOL operational noise through flight trajectory management," by F. H. Schmitz, et al. J AIRCRAFT 10:385-394, July, 1973.

HELICOPTER NOISE

"Test and evaluation of a quiet helicopter configuration HH-43B," by M. A. Bowes. ACOUSTICAL SOC AM J 54:1214-1218, November, 1973.

"Unified approach to aerodynamic sound generation in the presence of solid boundaries," by M. Goldstein. ACOUSTICAL SOC AM J 56: 497-509, August, 1974.

HOSPITAL NOISE

"Building noise in a hospital: an experimental simulation," by M. Powell. ANN OCCUP HYG 16:77-79, April, 1973.

"For goodness sake—Let your patients sleep!" by D. A. Grant, et al. NURSING 4:54-57, November, 1974.

"Hospital noise." (letter). N ENGL J MED 290:522-523, February 28, 1974.

"Hospital noise." NURS DIGEST 2:61, May, 1974.

"Hospital noise may cause patient problems." J ENVIRON HEALTH 36:354, January-February, 1974.

"Hospital noises can turn frightened patients into terrified patients," by W. A. Nolen. AMER MED NEWS 17:13, January 7, 1974.

"Hospital tranquility starts with mechanical systems: Nash General Hospital, Rocky Mount, N.C." HEATING PIPING AIR CONDITIONING 45:24, February, 1973.

"Noise in hospital." (editorial). BR MED J 4:625, December 15, 1973.

"Noise in hospitals," by R. Taylor. HEALTH SOC SERV J 84:2770-2771, November 30, 1974.

"Noise pollution in hospitals is regarded as health hazard." OR REPORTER 8:3 plus, October, 1973.

"Patients blame staffs in hospital noise control report." NURSING TIMES 70:251, February 21, 1974.

HOSPITAL NOISE

" 'Quiet campaign' featured at Methodist Hospital of Indianapolis, Ind." HOSP TOP 51:15, February, 1973.

"Quiet campaign—a sound idea." MICH HOSP 10:24, August, 1974.

"Supervisor overseas: hospital noise," by J. Wakeley. SUPERV NURSE 5:50, June, 1974.

HOUSEHOLD NOISE

HUNTERS

HYGIENIC STUDIES

"Data on the hygienic evaluation of city noise," by S. A. Soldatkina, et al. GIG SANIT 38:16-20, March, 1973.

"Device for the integral hygienic evaluation of noises," by P. N. Chumak, et al. GIG TR PROF ZABOL 17:48-50, July, 1973.

"Hygienic characteristics of noise and an analysis of the disease incidence among workers engaged in the metalworking industry," by E. P. Orlovskaia, et al. VRACH DELO 7:121-125, July, 1973.

IMPACT NOISE

"The effects to temporary hearing loss with combined impact and steady state noise," by K. Yamamura, et al. JAP J HYG 28(6):517-521, February, 1974.

"A method for the assessment of impact noise with respect to injury to hearing," by A. M. Martin, et al. ANN OCCUP HYG 16:19-26, April, 1973.

"Noise-induced hearing loss: the energy principle for recurrent impact noise and noise exposure close to the recommended limits," by G. R. Atherley. ANN OCCUP HYG 16(2):183-194, August, 1973.

IMPULSE NOISE

"Audiometric and anatomical correlates of impulse noise exposure," by D. Henderson, et al. ARCH OTOLARYNGOL 99:62-66, January, 1974.

IMPULSE NOISE

"Cochlear pathology in monkeys exposed to impulse noise," by V. M. Jordan, et al. ACTA OTOLARYNGOL [Suppl] 16-30, 1973.

"Deafening effects of impulse noise on the rhesus monkey." ACTA OTOLARYNGOL [Suppl] 1-44, 1973.

"Effect of impulse noise on workers' hearing," by L. I. Maksimova, et al. GIG SANIT 38:33-36, September, 1973.

"Impulse noise trauma. A study of histological susceptibility," by R. P. Hamernik, et al. ARCH OTOLARYNGOL 99:118-121, February, 1974.

"Interaction of continuous and impulse noise: audiometric and histological effects," by R. P. Hamernik, et al. J ACOUST SOC AM 55: 117-121, January, 1974.

"The relationship between permanent threshold shift and the loss of hair cells in monkeys exposed to impulse noise," by M. Pinheiro, et al. ACTA OTOLARYNGOL [Suppl] 31-40, 1973.

"Susceptibility to damage from impulse noise; chinchilla versus man or monkey," by G. A. Luz, et al. ACOUSTICAL SOC AM J 54:1750-1754, December, 1973.

INDUSTRIAL HYGIENE

"Hygienic working conditions and state of health of operators of track maintenance and repair machines," by E. I. Gol'Dman, et al. GIG TR PROF ZABOL 17(19):45-46, 1973.

INDUSTRIAL MEDICINE

"Industrial medicine. Mass screening of workers exposed to noise," (proceedings), by K. Humperdinck, et al. HEFTE UNFALLHEILKD 114:193-197, 1973.

INDUSTRIAL NOISE

"Characteristics of the cardiovascular system of adolescent workers subjected to the action of stable industrial noise," by E. A. Gel'tishcheva. GIG TR PROF ZABOL 16:29-33, July, 1973.

"Controlling industrial noise; acoustic materials and enclosures," by C. H. Wick. MANUF ENG & MGT 70:30-33, June, 1973.

"Controlling industrial noise; administrative controls and hearing protection," by C. H. Wick. MANUF ENG & MGT 71:32-35, July, 1973.

"Controlling noise at work [Great Britain]," by G. R. C. Atherley, et al. LABOR RESEARCH 63:150-154, July, 1974.

"Effect of different parameters of industrial noise on the auditory analyzer and the central nervous system of adolescent workers," by E. A. Gel'tishcheva. GIG TR PROF ZABOL 17:5-9, July, 1973.

"Effect of industrial noise and ototoxic antibiotics on cochlear function," by E. Krochmalska. ACTA OTOLARYNGOL 77:44-50, January-February, 1974.

"Effect of industrial noise on the hearing organ following conservative surgery of the middle ear," by W. Sulkowski, et al. OTOLARYNGOL POL 27:617-624, 1973.

"Effectiveness of different ear protectors in protecting the employee from over exposure in industrial environments," by J. E. Stephenson, et al. J ACOUST SOC AM 54(1):301, 1973.

"Effects of industrial noise upon the hearing organs following radical surgery of the middle ear," by J. Laciak, et al. OTOLARYNGOL POL 27:485-491, 1973.

"Getting noise immunity in industrial controls," by H. M. Schlicke, et al. IEEE SPECTRUM 10:30-35, June, 1973.

"Individual hearing protection—survey of Berlin's noisy factories," by P. Moch. ZENTRALBL ARBEITSMED 23:33-38, February, 1973.

"Industrial noise and hearing loss," by C. O. Istre, Jr, et al. J LA STATE MED SOC 126:5-7, January, 1974.

"Industrial noise and vibration in sewing industry enterprises and an evaluation of measures to decrease them," by V. F. Rudenko, et al. GIG TR PROF ZABOL 17:36-38, July, 1973.

"Modifications of epinephrine, norepinephrine, blood lipid fractions and the cardiovascular system produced by noise in an industrial medium," by G. A. Ortiz, et al. HORM RES 5:57-64, January, 1974.

"Morphologic changes in the nerve cells of the rabbit brain caused by industrial noise," by J. Tarmas, et al. FOLIA MORPHOL 33:5-12, 1974.

"Noise is causing an industrial headache," by D. W. Austin. MED TIMES 102:60-62, March, 1974.

"Noise of industrial enterprises and its effect on the population of Krivoi Rog," by N. M. Paran'ko, et al. GIG SANIT 37:98-99, July, 1972.

"Noise propagation in cellular urban and industrial spaces," by H. G. Davies, et al. ACOUSTICAL SOC AM J 54:1565-1570, December, 1973.

"Protection of residential areas from industrial noise," by P. Koltzsch, et al. Z GESAMTE HYG 19:331-337, May, 1973.

"Psychological effects of exposure to high industrial noise: a field study," by E. Gulian. J ACOUST SOC AM 55:Suppl:68, 1974.

"Techniques for industrial noise measurement," by R. A. Boole. PLANT ENG 28:105-107, February 7, 1974.

INDUSTRIAL NOISE: AEROPLANE FACTORY

INDUSTRIAL NOISE: CONSTRUCTION

"Construction noise; its origin and effects," by A. S. Hersh. AM SOC C E PROC 100:433-448, September, 1974.

"Construction, transportation tabbed as noise target areas." IND W 182: 18-19, July 1, 1974.

"New York State construction noise survey," by D. A. Driscoll, et al.

INDUSTRIAL NOISE: CONSTRUCTION

J ACOUST SOC AM 55:Suppl:37, 1974.

"Occupational hygiene and health status of concrete workers in construction of large hydroelectric power plants," by G. N. Metlyaev. GIG TR PROF ZABOL 17(12):8-11, 1973.

"Sound levels recommended for mobile construction equipment." AUTOMOTIVE ENG 81:23, June, 1973.

"Vibrations during construction operations," by J. F. Wiss. AM SOC C E PROC 100:239-246, September, 1974.

INDUSTRIAL NOISE: FLOUR

INDUSTRIAL NOISE: GLASS

INDUSTRIAL NOISE: NAIL FACTORY

INDUSTRIAL NOISE: ORE DRESSINY FACTORIES

INDUSTRIAL NOISE: RUBBER

INDUSTRIAL NOISE: SHOE-FACTORY

INDUSTRIAL NOISE: TEXTILE

INNER EAR
see also: Noise Research

"Diagnostic value of Bekesy comfortable loudness tracings," by J. Jerger, et al. ARCH OTOLARYNGOL 99:351-360, May, 1974.

"Effects of intense auditory stimulation: hearing losses and inner ear changes in the chinchilla," by I. M. Hunter-Duvar, et al. J ACOUST SOC AM 55:795-801, April, 1974.

"Effects of intense auditory stimulation: hearing losses and inner ear changes in the squirrel monkey," by I. M. Hunter-Duvar, et al. ACOUSTICAL SOC AM J 52:1181-1192; 54:1179-1183 pt 2 October, 1972, November, 1973.

INNER EAR

"Noise-induced inner ear damage in newborn and adult guinea pig," by S. A. Falk, et al. LARYNGOSCOPE 84:444-453, March, 1974.

"Noise-induced reduction of inner-ear microphonic response: dependence on body temperature," by D. G. Drescher. SCIENCE 185:273-274, July 19, 1974.

"Temporary threshold shift from a toy cap gun; Bekesy technique," by L. Marshall, et al. J SPEECH & HEARING DIS 39:163-168, May, 1974.

INSTRUMENTATION

IOPHENDYLATE

KANAMYCIN

LABORATORIES

"Noise pollution in the woodworking laboratory," by C. A. Pinder. MAN/SOC/TECH 34:47-50, November, 1974.

LANDSCAPING

"The role of green plants in the prevention of air and noise pollution." by G. Zamfir. REV MED CHIR SOC MED NAT IASI 77:673-678, October-December, 1973.

"Stop noise with hedges," by C. E. Whitcomb, et al. HORTICULTURE 52:58-59, April, 1974.

LARYNX

LAWS AND LEGISLATION

"California court bars class action; suit to recover damages for aircraft noise." AVIATION W 101:35, October 7, 1974.

"City of Burbank v. Lockheed Air Terminal, Inc. (93 Sup Ct 1854): federal preemption of aircraft noise regulations and the future of proprietary restrictions." NYU RES L & SOC CHANGE 4:99-113, Winter, 1974.

"Community noise control [South Florida]," by S. E. Dunn. FLA ENVIRONMENTAL & URBAN ISSUES 1:7-10, October-November, 1973.

"Construction, transportation tabbed as noise target areas." IND W 182: 18-19, July 1, 1974.

"Environmental law—aircraft noise control—use of local police powers to impose curfews on air flights is pre-empted by the federal aviation act of 1958 as amended by the noise control act of 1972." RUTGERS CAMDEN L J 5:566-584, Spring, 1974.

"Environmental law—aircraft noise regulation—federal pre-emption." NY L F 20:165-176, Summer, 1974.

"Environmental law—federal regulation of aircraft noise under federal aviation act precludes local police power noise restrictions." BC IND & COM L R 15:848-862, April, 1974.

"Environmental law—federal regulation under federal aviation act and noise control act preempts the field of airport and aircraft noise control rendering local airport curfews invalid." KAN L REV 22: 319-336, Winter, 1974.

"Environmental law: the noise control act of 1972." OKLA L REV 27: 55-62, Winter, 1974.

"EPA proposes rail noise standards." AIR POLLUTION CONTROL ASSN J 24:881, September, 1974.

"EPA readies noise recommendations," by R. K. Ellingsworth. AVIATION W 100:32-33, March 25, 1974.

"EPA report lists targets for noise standards." MACHINE DESIGN 46: 6, July 25, 1974.

"Expensive sound of silence." BUS W 28, July 20, 1974.

"Federal pre-emption and airport noise control." URBAN L ANN 8: 229-239, 1974.

LAWS AND LEGISLATION

"Federal pre-emption—aviation noise control—the Federal aviation administration, monitored by the Environmental protection agency, has full control over aviation noise, pre-empting state and local control, including a municipal ordinance which imposed a curfew on certain jet take-offs during certain night-time hours." J AIR L 40: 341-349, Spring, 1974.

"Federal preemption in airport noise abatement regulation: allocation of federal and state power." MAINE L REV 26:321-344, 1974.

"Fragmented noise control sought; EPA to press for noise funding scheme," by C. E. Schneider. AVIATION W 99:19-20, July 9; 32-33, July 16, 1973.

"Job safety-national compliance pact." FOOD PROCESSING 35:10, June, 1974.

"Medico-legal aspects of noise," by R. Murray. OCCUP HEALTH 25: 55-59, February, 1973.

"Mexican laws regarding control of smoke dust and noise pollution and their effects upon the oil milling industry," by M. Castaneda. J AM OIL CHEM SOC 51(2):279A, 1974.

"Milwaukee adopts noise law for home air conditioners," AIR COND HEAT & REFRIG N 130:1 plus, September 3, 1973.

"Noise control act of 1972—congress acts to fill the gap in environmental legislation." MINN L REV 58:273-306, December 1973.

"Noise control in Oregon: government regulation and private remedies," by J. F. Deits. WILLAMETTE L J 198-216, Spring, 1974.

"Noise, man and law," by E. Hammelburg. ORL 35:363-370, 1973.

"Noise pollution: public needs vs. individual rights," by D. Kastan. WESTERN STATE L REV 185-216, June, 1974.

" 'Noisy' battle over OSHA regulations," by H. G. Unger. CAN BUS 46: 8, November, 1973.

LAWS AND LEGISLATION

"OSHA, EPA disagree on operator exposure limits for noise," by E. Tabaczuk. AIR COND HEAT & REFRIG N 133:1 plus, December 9, 1974.

"OSHA issues tunnel rules, wrestles with noise." ENGIN N 192:10, April 18, 1974.

"OSHA noise standards too lenient, says EPA." CHEM MKTG REP 206: 16, December 23, 1974.

"An overview of EPA's implementation of the Noise Control Act of 1972," by A. F. Meyer, Jr. J AIR POLLUT CONTROL ASSOC 24: 830-831, September, 1974.

"Property—eminent domain—even where no direct overflight occurs, aircraft noise and air pollution depriving property owners of the practical use and enjoyment of their land is a taking requiring compensation for the diminution of the land's market value." J URBAN L 52:636-648, Winter, 1974.

"State regulation of nontransportation noise: law and technology," by R. W. Findley, et al. SO CALIF L REV 48:209-317, November, 1974.

"State standards, regulations, and responsibilities in noise pollution control," by J. M. Tyler, et al. J AIR POLLUT CONTROL ASSOC 24: 130-135, February, 1974.

"Take the thunder out of the big rigs; mandate from the EPA," by F. E. Bryson. MACHINE DESIGN 46:24-26 plus, September 19, 1974.

"Toward the comprehensive abatement of noise pollution: recent federal and New York city noise control legislation." ECOLOGY L Q 4: 109-144, Winter, 1974.

"Trucks are going to be quieter." IND W 181:57, April 8, 1974.

"Vehicular noise regulation in Hawaii," by J. C. Burgess. ACOUSTICAL SOC AM J 56:905-910, September, 1974.

LAWS AND LEGISLATION

"Voluntary noise pacts likely to spread." IND W 182:23-25, August 19, 1974.

"What's behind the proposed truck noise regulations?" AUTOMOTIVE ENG 82:42-5 plus, July, 1974.

LEARNING

"Delayed learning of rats exposed to noise," by T. Mitsuya. JAP J HYG 28:324-339, August, 1973.

"The effects of intelligence quotient and extraneous stimulation on incidental learning," by R. Forehand, et al. J MENT DEFIC RES 17: 24-27, March, 1973.

LIBRARIES

"Noise in libraries: causes and control," by A. Oluwakuyide. SPECIAL LIBRARIES 65,1:28-31, January, 1974.

LIGHTNING

"Acoustic and vestibular defects in lightning survivors," by J. W. Wright, Jr., et al. LARYNGOSCOPE 84(8):1378-1387, August, 1974.

MACHINE DESIGN

"Noise control leads to better bearing designs." PRODUCT ENG 45:21, January, 1974.

"Optimizing vent silencers design for gas blowdowns," by J. Hawkins. PIPELINES & GAS J 201:68 plus, October, 1974.

MACHINERY NOISE

"Active machine tool controller requirements for noise attenuation," by E. E. Mitchell, et al. J ENG IND 96:261-267, February, 1974.

"Analysis of errors in measuring machine noise under free-field conditions," by G. Hubner. ACOUSTICAL SOC AM J 54:967-977, October, 1973.

"Analyzing the sound of trouble," by R. E. Herzog. MACHINE DESIGN 45:128-134, September 6, 1973.

"Application of constrained-layer damping to control noise in machine parts," by P. D. Emerson. J ENG IND 96:299-303, February, 1974.

"Barriers for noise control," by J. N. Macduff. MECH ENG 96:26-31, August, 1974.

"Bring noise to book." ENGINEER 237:23, September 20, 1973.

"Build-them-yourself kits to muffle machine noise." ENGINEER 238:23, April 11, 1974.

"Controlling equipment noise," by W. Murtland. RUBBER WORLD 168:65-66, July, 1973.

"Curbing noise with partial enclosures," by W. G. Phillips, et al. MACHINE DESIGN 46:107-110, April 4, 1974.

"Gear noise source identification and reduction," by R. F. MacWhorter. AM IND HYG ASSOC J 35(9):581-585, September, 1974.

"Gear pump noise," by R. C. Michel. FUELOIL & OIL HEAT 33:46 plus, July, 1974.

"How to rate noise sources." MACHINE DESIGN 46:152, May 16, 1974.

"If you can't beat noise, baffle it." ENGINEER 237:23, September 13, 1973.

"In-place machinery noise measurements," by C. E. Ebbing, et al. ASHRAE J 15:48-54, June, 1973.

"Looking for noise?" FACTORY 7:54-55, February, 1974.

"Machine noise is reduced by fitting sliding shutters." ENGINEER 238:23, June 6, 1974.

"Machinery noise control," by R. T. Booth. OCCUP HEALTH 26:52-53, February, 1974.

"Measures to combat noise must allow for the public good," by P. McCallum. ENGINEER 238:36-37, April 18, 1974.

"Metalworking noise is a pain in the ear," by B. D. Wakefield. IRON AGE 212:55-60, December 13, 1973.

"MTIRA turns up volume on tool noise research." ENGINEER 238-9, May 30, 1974.

"New offender; intermittent noise," by B. D. Wakefield. IRON AGE 213:46-48, April 29, 1974.

"Noise and vibration analysis of an impact forming machine," by A. E. M. Osman, et al. J ENG IND 96:233-240, February, 1974.

"Noise control: a common-sense approach," by R. L. Lowery. MECH ENG 95:26-31, June, 1973.

"Noise measurement and control," by R. K. Miller. BLDG SYSTEMS DESIGN 70:41-44, June, 1973.

"Noise measurement standards for machines *in situ*," by W. W. Lang. ACOUSTICAL SOC AM J 54:960-966, October, 1973.

"Noise; a pain in the ear." METALLURGIA & METAL FORMING 41:55, March, 1974.

"Outlook for *in-situ* measurement of noise from machines," by T. J. Schulta. ACOUSTICAL SOC AM J 54:982-984, October, 1973.

"Plastic tube silences machinery." MACHINE DESIGN 46:40, April 4, 1974.

"Quiet bagging; Vyon silencers aid in machinery noise reduction." COMP AIR MAG 79:8-9, July, 1974.

"Quieting of process machinery," by H. A. Winnerling. CHEM ENG PROG 69:96-99, June, 1973.

"Shut up those sudden noises." ENGINEER 237:23, November 29, 1973.

MACHINERY NOISE

"Silent cabins, silencers, machine enclosures." ENGINEER 239:18, September 12, 1974.

"Sound power measurements on large machinery installed indoors," by G. M. Diehl. COMP AIR MAG 79:8-12, January, 1974.

"Tale of two noise-suppression treatments." MACHINE DESIGN 45:12, November 1, 1973.

"What you must do about controlling noise." MOD MATERIALS HANDLING 29:44-49, February, 1974.

"Why noise reduction doesn't always work." MACHINE DESIGN 46: 132, May 2, 1974.

MACHINERY NOISE: ELECTRIC
see: Motors: Electric

MACHINERY NOISE: FOUNDRY
"Divide and conquer your noise problems," by W. M. Ihde. FOUNDRY 101:61-62, August, 1973.

"Understanding how noise affects hearing." FOUNDRY 101:79-80 plus, July, 1973.

MACHINERY NOISE: HYDRAULIC
"Stopping hydraulic system noise," by H. W. Wojda. PLANT ENG 27: 74-75, July 26, 1973.

MACHINERY NOISE: LATHES

MACHINERY NOISE: PAPER MAKING

MACHINERY NOISE: TEXTILE
"Guidelines for textile industry noise control," by J. R. Bailey, et al. J ENG IND 96:241-246, February, 1974.

"Take the whine out of drawtwisting, and boost yarn quality as a bonus." TEXTILE WORLD 123:38 plus, November, 1973.

MAN
see also: Behavior
Physiology
Psychology

"Analysis of evoked responses in man elicited by sinusoidally modulated noise," by M. Rodenburg, et al. AUDIOLOGY 11:283-293, September-December, 1972.

"The effect of acute noise exposure on the excretion of corticosteroids, adrenalin and noredrenalin in man," by A. Slob, et al. INT ARCH ARBEITSMED 31:225-235, July 10, 1973.

"Effect of noise on man," by A. Mann. HAREFUAH 83:387-388, November 1, 1972.

"Man and the environmental noise," by E. Wende, et al. INTERNIST 14:224-229, May, 1973.

"Noise, man and law," by E. Hammelburg. ORL 35:363-370, 1973.

"Some methodological problems in studying the action of noise on the body of man and animals," by L. I. Maksimova, et al. GIG SANIT 38:30-35, July, 1973.

MASKING NOISE
"The masking noise and its effect upon the human cortical evoked potential," by H. G. Ghueden. AUDIOLOGY 11:90-96, January-April, 1972.

"The masking of binaural beats of a pure sound with a differential sound," by R. Piazza. AUDIOLOGY 11:169-176, May-August, 1972.

METALLIC MERCURY

METHIONIN-S35

MICROPHONES
"Leakage path of room noise to the phonocardiographic microphones,"

MICROPHONES

by N. Suzumura, et al. JAP J MED ELECTRON 11:344-349, October, 1973.

"On the noise level of ears and microphone," by M. C. Killion. J ACOUST SOC AM 55:Suppl:41, 1974.

MIDDLE EAR

"Acoustic neurinomas presenting as middle ear tumors," by L. A. Storrs. LARYNGOSCOPE 84:1175-1180, July, 1974.

"Effect of industrial noise on the hearing organ following conservative surgery of the middle ear," by W. Sulkowski, et al. OTOLARYNGOL POL 27:617-624, 1973.

"Effects of industrial noise upon the hearing organs following radical surgery of the middle ear," by J. Laciak, et al. OTOLARYNGOL POL 27:485-491, 1973.

"Middle ear measurements," by G. T. Wolcott, et al. J MED ASSOC STATE ALA 43:496-498, February, 1974.

MINIBIKES

"Hearing loss due to minibikes," by R. P. Oppenheimer, et al. AM FAM PHYSICIAN 8:125, October, 1973.

MINING NOISE

"Designing safety into underground mining equipment; noise abatement," by C. Holvenstot. MIN CONG J 59:39-43, September, 1973.

"Effect of environment on the health of the mining personnel in copper ore mines of the Legnica-Glogow Copper Region," by I. Juzwiak, et al. POL TYG LEK 29:811-814, May 13, 1974.

"Effect of noise on the ear following tympanoplasty in miners employed underground," by S. Stawinski. OTOLARYNGOL POL 27:751-755, 1973.

MOTOR TRUCKS

"The exposure of truck drivers to noise," by B. H. Sharp. J ACOUST SOC AM 55(2):484, 1974.

MOTOR TRUCKS

"Next ten years in truck technology; noise." AUTOMOTIVE ENG 81: 37 plus, June, 1973.

"Noise and the truck drivers," by D. A. Tyler. AM IND HYG ASSOC J 34:345-349, August, 1973.

"Shielding cuts truck and bus noise." AUTOMOTIVE ENG 81:15-16, August, 1973.

"Take the thunder out of the big rigs: mandate from the EPA," by F. E. Bryson. MACHINE DESIGN 46:24-26 plus, September 19, 1974.

"Truck noise problem and what might be done about it," by R. F. Ringham. AUTOMOTIVE ENG 81:29-31 plus, April, 1973.

"Trucks are going to be quieter." IND W 181:57, April 8, 1974.

"Quiet refuse truck beats 75 decibels." FLEET OWNER 69:170, March, 1974.

"Quiet truck program progressing at International." AUTOMOTIVE IND 150:138 plus, April 1, 1974.

"Volvo designs a quiet tractor cab." AUTOMOTIVE ENG 81:68-71, September, 1973.

"What's behind the proposed truck noise regulations?" AUTOMOTIVE ENG 82:42-45 plus, July, 1974.

MOTORS: ELECTRIC

"Four leads slash capacitor noise." MACHINE DESIGN 46:47, August 8, 1974.

"Lowering equipment noise levels; questions and answers." ELEC CONSTR & MAINT 72:142, August, 1973.

"Noise considerations on large process unit drivers," by T. C. Again, et al. IEEE TRANS IND APPLICATIONS 10:296-304, March, 1974.

MUSCLES: SKELETAL

MUSIC
see also: Electronic Noise

"Effect of loud music on hearing." (editorial). W VA MED J 70:165-166, July, 1974.

"Electronic reverberation equipment in the Stockholm concert hall," by S. Dahlstedt. AUDIO ENG SOC J 22:627-631, October, 1974.

"Hearing damage from music; United Kingdom experience," by A. Burd. AUDIO ENG SOC J 22:524-527, September, 1974.

"New automatic noise-reduction system (ANRS)," by M. Yamazaki, et al. AUDIO ENG SOC J 21:445-449, July-August, 1973.

"Oranges save Britain from rock." MACHINE DESIGN 45:47, December 13, 1973.

" 'Piano killer' strikes a responsive chord in noise-filled Japan: slaying of woman, two daughters over music sparks debate in an overcrowded nation," by N. Pearlstine. WALL ST J 184:1 plus, December 3, 1974.

"A pilot investigation of noise hazards in recording studios," by G. W. Gibbs, et al. ANN OCCUP HYG 16(4):321-327, 1973.

"Temporary hearing losses in teenagers attending repeated rock-and-roll sessions," by R. F. Ulrich. ACTA OTO-LARYNGOL 77(1-2):51-55, 1974.

MUSICAL INSTRUMENTS
see also: Music

MUSICIANS
see: Music
Musical Instruments

NASA
see also: Aircraft Noise

NASA

"NASA JT8D refan program nears end; noise reductions achieved with refanned jet engines," by M. L. Yaffee. AVIATION W 101:46-47, July 22, 1974.

NEOMYCIN

NEONATAL

"Committee on Environmental Hazards. Noise pollution: neonatal aspects." PEDIATRICS 54(4):476-479, October, 1974.

"The effect of white noise on the somatosensory evoked response in sleeping newborn infants," by P. H. Wolff, et al. ELECTROENCEPHALOGR CLIN NEUROPHYSIOL 37:269-274, September, 1974.

"Noise levels in infant incubators (adverse effects?)," by G. Blennow, et al. PEDIATRICS 53:29-32, January, 1974.

"Responsiveness to simple and complex auditory stimuli in the human newborn," by G. Turkewitz, et al. DEV PSYCHOBIOL 5:7-19, 1972.

"Square-wave stimuli and neonatal auditory behavior: reply to Bench," (letter), by S. J. Hutt. J EXP CHILD PSYCHOL 16:530-533, December, 1973.

" 'Square-wave stimuli' and neonatal auditory behavior: some comments on Ashton (1971), Hutt et al. (1968) and Lenard et al. (1969)," by J. Bench. J EXP CHILD PSYCHOL 16:521-527, December, 1973.

"The use of alternated stimuli to reduce response decrement in the auditory testing of newborn infants," by D. Ling, et al. J SPEECH HEAR RES 14:531-534, September, 1971.

NERVOUS SYSTEM
see also: Noise Research

"Analysis of central nervous system involvement in the microwave auditory effect," by E. M. Taylor, et al. BRAIN RES 74:201-208, July 12, 1974.

NERVOUS SYSTEM

"Effect of different parameters of industrial noise on the auditory analyzer and the central nervous system of adolescent workers," by E. A. Gel'tishcheva. GIG TR PROF ZABOL 17:5-9, July, 1973.

NITROUS OXIDE

NOISE: AUSTRIA

NOISE: CANADA

NOISE: GERMANY
"Individual hearing protection—survey of Berlin's noisy factories," by P. Moch. ZENTRALBL ARBEITSMED 23:33-38, February, 1973.

NOISE: ITALY

NOISE: JAPAN
" 'Piano killer' strikes a responsive chord in noise-filled Japan: slaying of woman, two daughters over music sparks debate in an overcrowded nation," by N. Pearlstine. WALL ST J 184:1 plus, December 3, 1974.

NOISE: MEXICO
"Measuring procedure for urban noise in the center of Mexico City," by F. Groenwold, et al. J ACOUST SOC AM 55(2):465, 1974.

"Mexican laws regarding control of smoke dust and noise pollution and their effects upon the oil milling industry," by M. Castaneda. J AM OIL CHEM SOC 51(2):279A, 1974.

NOISE: POLAND

NOISE: SWEDEN
"Aircraft noise now on-line to Stockholm's new pollution monitoring network." ATMOSPHERIC ENVIRONMENT 7:Suppl:ii-iii, December, 1973.

NOISE: UNITED KINGDOM
"Election kills noise bill (Environment Protection Bill)." MEL MAKER 49:5, February 16, 1974.

NOISE: UNITED KINGDOM

"High noise limits will hit British truck men." ENGINEER 237:1, August 16, 1973.

"Oranges save Britain from rock." MACHINE DESIGN 45:47, December 13, 1973.

"School children in London; noise survey." HEALTH VISIT 45:274-277, September, 1972.

NOISE: UNITED STATES

CALIFORNIA

"California court bars class action; suit to recover damages for aircraft noise." AVIATION W 101:35, October 7, 1974.

"City of Burbank v. Lockheed Air Terminal Inc. (93 Sup Ct 1854); federal preemption of aircraft noise regulation and the future of proprietary restrictions." NYU REV L & SOC CHANGE 4:99-113, Winter, 1974.

CONNECTICUT

"Noise control in Connecticut." DIESEL EQUIP SUPT 52:46, September, 1974.

FLORIDA

"Community noise control [South Florida]," by S. E. Dunn. FLA ENVIRONMENTAL & URBAN ISSUES 1:7-10, October-November, 1973.

HAWAII

"Vehicular noise regulation in Hawaii," by J. C. Burgess. ACOUSTICAL SOC AM J 56:905-910, September, 1974.

ILLINOIS

"Aircraft noise abatement via annex 16 of the Chicago convention—a viable alternative," by S. S. Kalsi. TEX INT L J 9:1-18, Winter, 1974.

"Background noise study in Chicago," by C. Caccavari, et al. AIR POLLUTION CONTROL ASSN J 24:240-244, March, 1974.

NOISE: UNITED STATES

ILLINOIS

"Big noise from Illinois," by F. M. H. Gregory. HOT ROD 27:54-55, January, 1974.

"Positive progress announced on Illinois noise problem." HOT ROD 27: 32, June, 1974.

INDIANA

" 'Quiet campaign' featured at Methodist Hospital of Indianapolis, Ind." HOSP TOP 51:15, February, 1973.

LOUISIANA

MASSACHUSETTS

"Community noise survey of Medford, Massachusetts," by J. E. Wesler. J ACOUST SOC AM 54:985-995, October, 1973.

NEW YORK

"New York State construction noise survey," by D. A. Driscoll, et al. J ACOUST SOC AM 55:Suppl:37, 1974.

"Toward the comprehensive abatement of noise pollution: recent federal and New York city noise control legislation." ECOLOGY L Q 4: 109-144, Winter, 1974.

OREGON

"Noise control in Oregon: government regulation and private remedies." WILLAMETTE L J 10:198-216, Spring, 1974.

WISCONSIN

"Milwaukee adopts noise law for home air conditioners." AIR COND HEAT & REFRIG N 130:1 plus, September 3, 1973.

WASHINGTON

"Aircraft noise induced vibration in fifteen residences near Seattle-Tacoma International Airport," by S. M. Cant, et al. AM IND HYG ASSOC J 34:463-468, October, 1973.

NOISE: USSR

"Active machine tool controller requirements for noise attenuation," by E. E. Mitchell, et al. J ENG IND 96:261-267, February, 1974.

"Aircraft noise abatement via annex 16 of the Chicago convention—a viable alternative," by S. S. Kalsi. TEX INT L J 9:1-18, Winter, 1974.

"Build-them-yourself kits to muffle machine noise." ENGINEER 238-23, April 11, 1974.

"Burwen dynamic noise filter," by B. Whyte. AUDIO 58:10 plus, February, 1974.

"A cost-effective method of evaluating aircraft noise-abatement options [San Antonio international airport]," by S. R. Goldberg. TEX BUS R 47:284-287, December, 1973.

"Designing safety into underground mining equipment; noise abatement," by C. Holvenstot. MIN CONG J 59:39-43, September, 1973.

"Dolby B-type noise reduction system," by R. Berkovitz, et al. AUDIO 57:15-16, September; 33-36, October, 1973.

"Federal preemption in airport noise abatement regulation: allocation of federal and state power." MAINE L REV 26:321-344, 1974.

"Kinematic sound screen; unique solution to highway noise abatement," by J. B. Hauskins, Jr. AM SOC C E PROC 100:169-178, February, 1974.

"New interpretation of noise reduction by matching," by Y. Nezer. IEEE PROC 62:404-406, March, 1974.

"Noise abatement," by P. Kelsey. AIR COND HEAT & REFRIG N 132: 1 plus, July 15, 1974.

"Noise abatement, problem and progress, symposium." DIESEL EQUIP SUPT 52:42-44, June, 1974.

"Noise abatement program," by M. H. Miller. J ACOUST SOC AM 54

NOISE ABATEMENT

(1):288, 1973.

"Noise: meter men move in." MEL MAKER 49:5, January 12, 1974.

"Reducing noise in food plants," by R. K. Miller. FOOD ENG 46:75-76, February, 1974.

"Spaced can feeding reduces noise from 100 to 85 db [Shasta]," by E. Lane, et al. FOOD PROCESSING 34:38-39, November, 1973.

"Toward the comprehensive abatement of noise pollution: recent federal and New York city noise control legislation." ECOLOGY L Q 4: 109-144, Winter, 1974.

"Why noise reduction doesn't always work." MACHINE DESIGN 46: 132, May 2, 1974.

NOISE LEVELS

"Audible noise levels of oxygen masks operating on venturi principle," by J. M. Leigh. BR MED J 4:652, December 15, 1973.

"Calculating noise levels of fans; nomograph," by F. Caplan. PLANT ENG 28:88-89, January 24, 1974.

"Effect of ambient illumination on noise level of groups," by M. Sanders, et al. J APP PSYCHOL 59:527-528, August, 1974.

"Effect of an interstate highway on urban area noise levels," by J. E. Heer, Jr., et al. PUB WORKS 105:54-58, January, 1974.

"On the noise level of ears and microphones," by M. C. Killion. J A-COUST SOC AM 55:Suppl:41, 1974.

NOISE MEASUREMENT
see also: Noise Measurement Devices

"Analysis of errors in measuring machine noise under free-field conditions," by G. Hubner. ACOUSTICAL SOC AM J 54:967-977, October, 1973.

"Automatic urban noise monitoring and analysis system," by J. E. K. Foreman, et al. ACOUSTICAL SOC AM J 55:1358-1359, June, 1974.

"Comparison of inside and outside noise measurements in various urban environments," by D. E. Bishop. J ACOUST SOC AM 55(2):465, 1974.

"In-place machinery noise measurements," by C. E. Ebbing, et al. ASHRAE J 15:48-54, June, 1973.

"The measurement of noise from moving vehicles," by E. L. Hixson, et al. J ACOUST SOC AM 54(1):332, 1973.

"Measuring procedure for urban noise in the center of Mexico City," by F. Groenwold, et al. J ACOUST SOC AM 55(2):465, 1974.

"Noise measurement and control," by R. K. Miller. BLDG SYSTEMS DESIGN 70:41-44, June, 1973.

"Noise measurement standards for machines *in situ*," by W. W. Lang. SOC AM J 54:960-966, October, 1973.

"Noise measurement techniques," by J. Donovan. ASTM STAND N 2:17-31 plus, May, 1974.

"Noise measurements in a university: an open-ended student experiment," by A. A. Silvidi, et al. AMERICAN JOURNAL OF PHYSICS 41,7:909-913, July, 1973.

"Outlook for *in-situ* measurement of noise from machines," by T. J. Schultz. ACOUSTICAL SOC AM J 54:982-984, October, 1974.

"Relation of noise measurements to temporary threshold shift in snowmobile users," by R. B. Chaney, Jr., et al. J ACOUST SOC AM 54:1219-1223, November, 1973.

"Sounds around us [table]." AIR COND HEAT & REFRIG N 133:39, November 4, 1974.

NOISE MEASUREMENT

"Sound-level measurements in the community," by D. S. Allen. AIR COND HEAT & REFRIG N 131:30-32, March 4, 1974.

"Techniques for industrial noise measurement," by R. A. Boole. PLANT ENG 28:105-107, February 7, 1974.

"Understanding decibels," by G. Board. POP ELECTR 5:94-95, April, 1974.

NOISE MEASUREMENT DEVICES
see also: Noise Measurement

"Another way to measure noise [dosimeters]." FACTORY 7:55, February, 1974.

"Home builder group rents out noise measuring instruments," by K. MacDonald. AIR COND HEAT & REFRIG N 130:37, October 29, 1973.

"A noise pollution level instrument," by J. A. Hamburg. REV SCI INSTRUM 44:1618-1620, November, 1973.

"Personal noise dosimetry in refinery and chemical plants," by A. H. Diserens. J OCCUP MED 16:255-257, April, 1974.

"Portable measuring device for the equivalent gauge of permanent noise," by W. Liebig. Z GESAMTE HUG 20:1-4, January, 1974.

NOISE RESEARCH
"Acoustic trauma after double exposure in mammals," by A. Pye. AUDIOLOGY 13(4):320-325, 1974.

"Acute effects of ethanol on spontaneous and auditory evoked electrical activity in cat brain," by R. G. Perrin, et al. ELECTROENCEPHALOGR CLIN NEUROPHYSIOL 36:19-31, January, 1974.

"Adaptation for sound localization in the ear and brainstem of mammals," by R. B. Masterton. FED PROC 33:1904-1910, August, 1974.

"Alterations in morphine-induced analgesia in mice exposed to pain, light or sound," by M. W. Stevens, et al. ARCH INT PHARMACODYN THER 206:66-75, November, 1973.

"Analysis of information-bearing elements in complex sounds by auditory neurons of bats," by N. Suga. AUDIOLOGY 11:58-72, January-April, 1972.

"Animal and human tolerance of high-dose intramuscular therapy with spectinomycin," by E. Novak, et al. J INFECT DIS 130:50-55, July, 1974.

"Azure B-RNA changes in the adrenal and cerebral cortex of rats exposed to intense noise," by A. Anthony. FED PROC 32:2093-2097, November, 1973.

"Barth's myochordotonal organ as an acoustic sensor in the ghost crab, Ocypode," by K. Horch. EXPERIENTIA 30:630-631, June 15, 1974.

"Blasting noise research project." PIT & QUARRY 66:26-27, March, 1974.

"Change in the reactivity of the terminal vessels of the rat brain under the action of stable noise," by S. V. Alekseev, et al. GIG TR PROF ZABOL 17:18-21, July, 1973.

"Cochlear electrical activity in noise-induced hearing loss behavioral and electrophysiological studies in primates," by J. E. Pugh, Jr., et al. ARCH OTOLARYNGOL 100:36-40, July, 1974.

"Cochlear pathology in monkeys exposed to impulse noise," by V. M. Jordan, et al. ACTA OTOLARYNGOL [Suppl] 16-30, 1973.

"Convulsive seizures in autostimulation during the period of sensitivity to audiogenic seizure in DBA-2 mice," by P. Cazala, et al. C R ACAD SCI [D] 278(22):2811-2814, May 27, 1974.

"Combination effects of carbon tetrachloride and noise on GPT and LAP activity in serum of rats," by G. Wagner, et al. Z GESAMTE HYG

19:862-863, 1973.

"Deafening effects of impulse noise on the rhesus monkey." ACTA OTO-LARYNGOL [Suppl] 1-44, 1973.

"Decay of temporary threshold shift in noise; monaural chinchillas," by J. H. Mills, et al. J SPEECH & HEARING RES 16:267-270, June, 1973.

"Delayed learning of rats exposed to noise," by T. Mitsuya. JAP J HYG 28:324-339, August, 1973.

"Discharge patterns of single fibers in the pigeon auditory nerve," by M. B. Sachs, et al. BRAIN RES 70:431-447, April 26, 1974.

"Dorsal cochlear nucleus of the chinchilla: excitation by contralateral sound," by T. E. Mast. BRAIN RES 62:61-70, November 9, 1973.

"Effect of D-amphetamine sulfate on susceptibiltiy to audiogenic seizures in DBA-2J mice," by J. M. Graham, Jr., et al. BEHAV BIOL 10:183-190, February, 1974.

"Effect of intensive noise on the microcirculation in the brain of experimental animals," by S. V. Alekseev, et al. GIG TR PROF ZABOL 16:24-26, July, 1972.

"Effect of priming and testing for audiogenic seizures in BALB-c mice as a function of stimulus intensity," by C. S. Chen, et al. EXPERIENTIA 30:153, February 15, 1974.

"The effect of vibration and noise on development of inflammatory reaction in rats," by J. Billewicz-Stankiewicz, et al. ACTA PHYSIOL POL 25:235-240, May, 1974.

"Effects of intense auditory stimulation: hearing losses and inner ear changes in the chinchilla," by I. M. Hunter-Duvar, et al. J ACOUST SOC AM 55:795-801, April, 1974.

"Effects of intense auditory stimulation: hearing losses and inner ear changes in the squirrel monkey. II.," by I. M. Hunter-Duvar, et al.

J ACOUST SOC AM 54:1179-1183, November, 1973.

"Effects of low-frequency vibration and noise on conditioned avoidance reaction in rats," by J. Billewicz-Stankiewicz, et al. ACTA PHYSIOL POL 25(4):307-312, July-August, 1974.

"The effects of noise level and elevated ambient temperatures upon selected reproductive traits in female Swiss-Webster mice," by H. B. Zakem, et al. LAB ANIM SIC 24:469-475, June, 1974.

"Effects of noise pollution on animal behavior," by W. E. Brewer. CLIN TOXICOL 7:179-189, April, 1974.

"Engine-over-the-wing noise research," by M. Reshotko, et al. J AIRCRAFT 11:195-196, April, 1974.

"A gene controlling bell- and photically-induced ovulation in mice," by B. E. Eleftherious, et al. J REPROD FERTIL 38:41-47, May, 1974.

"High-intensity ultrasonic sound: a better rat," by B. J. Morley, et al. PSYCHOL REPT 35:152-154, August, 1974.

"Histochemical activity of succinate dehydrogenase in guinea pig cochlea after impulse stimulation," by H Guttmacher, et al. ACTA OTOLARYNGOL 76:323-327, November, 1973.

"Lesions in the septal nuclei of the rat raise mean systemic arterial pressure and prevent the development of sound-withdrawal hypertension," by J. F. Marwood, et al. J PHARM PHARMACOL 25:614-620, August, 1973.

"Localization of acoustic stimulation in fishes and amphibia," by E. Schwartz. FORTSCHR ZOOL 21:121-135, 1973.

"Measurement of 'instantaneous' carrier frequency of bat pulses," (letter)," by P. J. Kindlmann, et al. J ACOUST SOC AM 54:1380-1382, November, 1973.

"Modification of the rat's startle reaction by an antecedent change in the acoustic environment," by C. L. Stitt, et al. J COMP PHYSIOL PSY-

CHOL 86:826-836, May, 1974.

"Monkeys agree-noise is upsetting." MED TIMES 102:71-72 plus, March, 1974.

"Monosynaptic projections from the pontine reticular formation to the 3rd nucleus in the cat," by S. M. Highstein, et al. BRAIN RES 75: 340-344, July 26, 1974.

"Morphologic changes in the nerve cells of the rabbit brain caused by industrial noise," by J. Tarmas, et al. FOLIA MORPHOL 33:5-12, 1974.

"Morphological differentiation of nerve cells in rats of two strains with different genetically determined reactions to sound," by I. Y. Raushenbakh, et al. SOV J DEV BIOL 3:133-138, March-April, 1972.

"Morphology and function of the dorsal sound producing scales in the tail of Teratoscincus scincus (Reptilia; Gekkonidae)," by U. Hiller. J MORPHOL 144(1):199-130, September, 1974.

"Noise-induced inner ear damage in newborn and adult guinea pigs," by S. A. Falk, et al. LARYNGOSCOPE 84:444-453, March, 1974.

"Noise-induced threshold shift in the parakeet (Melopsittacus undulatus)," by J. Saunders, et al. PROC NATL ACAD SCI USA 71: 1962-1965, May, 1974.

"Noise, vitamin A deficiency, and emotional behavior in rats," by A. Wells, et al. PERCEPT MOT SKILLS 38:392-394, April, 1974.

"Other medical views and research on noise." MED TIMES 102:84-86 plus, March, 1974.

"Polysensory responses and sensory interaction in pulvinar and related postero-lateral thalamic nuclei in cat," by C. C. Huang, et al. ELECTROENCEPHALOGR CLIN NEUROPHYSIOL 34:265-280, March, 1973.

"Postnatal development of provoked activity in the superior lateral olive in the cat by stimulation by sound," by R. Romand, et al. J PHYSIOL 66:303-315, September, 1973.

"Priming for audiogenic seizures in mice: influence of postpriming auditory environment," by G. R. Bock, et al. EXP NEUROL 42:700-702, March, 1974.

"Production of calibrated sound pressures at the tympanic membrane of the guinea pig," by H. Wagner, et al. ARCH OTORHINOLARYNGOL 206:283-292, June 18, 1974.

"Protection from lethal audiogenic seizures in mice by physical restraint of movement," by J. F. Willott. EXP NEUROL 43:359-368, May, 1974.

"The relation between temporary threshold shift and permanent threshold shift in rhesus monkeys exposed to impulse noise," by G. A. Luz, et al. ACTA OTOLARYNGOL [Suppl] 1-15, 1973.

"The relationship between permanent threshold shift and the loss of hair cells in monkeys exposed to impulse noise," by M. Pinheiro, et al. ACTA OTOLARYNGOL [Suppl] 31-40, 1973.

"Research on the unitary discharges of cells of the anterior sigmoid gyrus after acoustic stimulation in normal animals and in animals whose acoustic areas have been removed bilaterally," by O. Sager, et al. REV ROUM NEUROL 10:61-73, 1973.

"Response of the amphibian papilla nerve in the toad bufomarinus," (proceedings), by H. Oyama. J PHYSIOL SOC JAP 35:534-535, August-September, 1973.

"The response of the swim bladder of the goldfish (Carassium auratus) to acoustic stimuli," by A. N. Pipper. J EXP BIOL 60:295-304, April, 1974.

"Responses of neurons in auditory cortex of the macaque monkey to monaural and binaural stimulation," by J. F. Brugge, et al. J NEUROPHYSIOL 36:1138-1158, November, 1973.

"Role of auditory cortex in sound localization: a comparative ablation study of hedgehog and bushbaby," by R. Ravizza, et al. FED PROC 33:1917-1919, August, 1974.

"Role of sound, transmitted through the air or a substrate, in the communications of social insects," by E. K. Es'kov. ZH OBSHCH BIOL 34:861-871, November-December, 1973.

"Sensitization of the rat startle response by noise," by M. Davis. J COMP PHYSIOL PSYCHOL 87(3):571-581, September, 1974.

"Shock tolerance in rats as a function of white noise," by M. Cunningham, et al. PSYCHOL REP 34:711-713, June, 1974.

"Signal-to-noise ratio as a predictor of startle amplitude and habituation in the rat," by M. Davis. J COMP PHYSIOL PSYCHOL 86:812-825, May, 1974.

"Singing muscles in a katydid," (letter), by J. W. Pringle. NATURE 250: 442, August 2, 1974.

"Some methodological problems in studying the action of noise on the body of man and animals," by L. I. Maksimova, et al. GIG SANIT 38:30-35, July, 1973.

"Sound and sound emission apparatus in puerulus and postpuerulus of the western rock lobster (Panulirus longipes)," by V. B. Meyer-Rochow, et al. J EXP ZOOL 189:283-289, August, 1974.

"Sound deprivation causes hypertension in rats," by M. F. Lockett, et al. FED PROC 32:2111-2114, November, 1973.

"Sound production in scolytidae: 'rivalry' behaviour of male Dendroctonus beetles," by J. A. Rudinsky, et al. J INSECT PHYSIOL 20: 1219-1230, July, 1974.

"Superior vestibular and 'singular nerve' section—animal and clinical studies," by H. Silverstein, et al. LARYNGOSCOPE 83:1414-1432, September, 1973.

"Susceptibility to damage from impulse noise: chinchilla versus man or monkey," by G. A. Luz, et al. ACOUSTICAL SOC AM J 54:1750-1754, December, 1973.

"Symposium: new data for noise standards. IV. The physiological effects of priming for audiogenic seizures in mice," by J. C. Saunders. LARYNGOSCOPE 84:750-758, May, 1974.

"Threshold shifts produced by exposure to noise in chinchillas with noise-induced hearing losses," by J. H. Mills. J SPEECH HEAR RES 16:700-708, December, 1973.

"Vestibular and auditory cortical projection in the guinea pig (Cavia porcellus)," by L. M. Odkvist, et al. EXP BRAIN RES 18:279-286, October 26, 1973.

NOISE STANDARDS

"Big noises are being heard as industry considers the cost," by C. Beatson. ENGINEER 238:38-39 plus, February 7, 1974.

"Can we achieve a quieter environment?" by K. S. Oliphant. ASTM STAND N 2:8-13, May, 1974.

"Consumer product noise reduction," by R. S. Musa. J ENVIRONMENTAL SCI 17:9-11, July, 1974.

"Criteria for a recommended standard—occupational exposure to noise. I. Recommendations for a noise standard," by H. M. Utidjian. J OCCUP MED 16:33-37, January, 1974.

"Downing the plant's din." CHEM ENG 80:30 plus, December 24, 1973.

"Environmental noise management," by D. P. Loucks, et al. AM SOC C E PROC 99'813-829, December, 1973.

"Environmental noise; status of the Agency programs." ASTM STAND N 2:32 plus, May, 1974.

"EPA proposes rail noise standards." AIR POLLUTION CONTROL ASSN J 24:881, September, 1974.

"EPA report lists targets for noise standards." MACHINE DESIGN 46: 6, July 25, 1974.

"Establishment of noise level standards in administrative and public buildings," by V. A. Tokarev, et al. GIG TR PROF ZABOL 0(8): 13-16, August, 1974.

"Europe seeks common solutions to problems of emissions and noise." AUTOMOTIVE ENG 81:25-31, March, 1973.

"First result of the noise control act; how noise affects people," by R. A. Jacobson. MACHINE DESIGN 45:132-136, October 18, 1973.

"Guidelines for noise control." AM DYESTUFF REP 62:59-63, July; 35-37 plus, August, 1973.

"Guidelines for textile industry noise control," by J. R. Bailey, et al. J ENG IND 96:241-246, February, 1974.

"High noise limits will hit British truck men." ENGINEER 237:1, August 16, 1973.

"Instrumentation for measuring and analyzing noise," by T. T. Weissenburger. PLANT ENG 27:80-84, November 1, 1973.

"Interdisciplinary plant-noise control," by A. Thumann. CHEM ENG 81: 120 plus, August 19, 1974.

"International standardization for noise." ASTM STAND N 2:50, May, 1974.

"Measures to combat noise must allow for the public good," by P. McCallum. ENGINEER 238:36-37, April 18, 1974.

"Mechanical equipment noise and vibration control," by L. F. Yerges. HEATING PIPING 45:61-66, July, 1973.

"Need for noise control." METALLURGIA & METAL FORMING 40: 270, September, 1973.

"New noise exposure level may become OSHA standard." AUTOMATION 20:12, October, 1973.

"Noise and Uncle," by W. L. Clevenger. AUDIO ENG SOC J 21:724-726, November, 1973.

"Noise control: an act where purchasing can take the lead," by C. H. Deutsch. PURCHASING 75:19 plus, December 18, 1973.

"Noise control and civil engineering," by E. M. Krokosky, et al. CIVIL ENG 44:45-49, May, 1974.

"Noise-control design for process plants," by S. C. Lou. CHEM ENG 80:77-82, November 26, 1973.

"Noise control office near action after years of study." IND W 181:18 plus, June 3, 1974.

"Noise control versus shock and vibration engineering," by C. T. Morrow. ACOUSTICAL SOC AM J 55:695-699, April, 1974.

"Noise; future targets," by G. M. Lilley. AERONAUTICAL J 78:459-463, October, 1974.

"OSHA noise; guilty! what did you say?" IRON AGE 212:36, November 22, 1973.

"OSHA noise standards too lenient, says EPA." CHEM MKTG REP 206:16, December 23, 1974.

"Predict plant noise problems," by R. S. Norman. HYDROCARBON PROCESS 52:89-91, October, 1973.

"Quieter please!" by I. Berkovitch. ENGINEERING 214:392-394, May, 1974.

"Social impact of aricraft noise," by A. Alexandre. TRAFFIC Q 28:371-388, July, 1974.

"State standards, regulations, and responsibilities in noise pollution control," by J. M. Tyler, et al. AIR POLLUTION CONTROL ASSN J 24:130-135, February, 1974.

"Surface transportation noise and its control," by J. E. Wesler. AIR POLLUTION CONTROL ASSN J 23:701-703, August, 1973.

"Symposium: new data for noise standards. I. New data for noise standards," by D. Henderson, et al. LARYNGOSCOPE 84:714-721, May, 1974.

--IV. The physiological effects of priming for audiogenic seizures in mice," by J. C. Saunders. LARYNGOSCOPE 84:750-756, May, 1974.

"U.S. noise standards." LABOUR GAZ 74:692, October, 1974.

"What you must do about controlling noise." MOD MATERIALS HANDLING 29:44-49, February, 1974.

"You can learn to live with noise control," by W. A. Bradley. POWER 117:66-67, September, 1973.

NOISE STUDIES

"Background noise study in Chicago," by C. Caccavari, et al. AIR POLLUTION CONTROL ASSN J 24:240-244, March, 1974.

"Fallacies of silence," by H. Carruth. HUDSON R 26:462-470, Autumn, 1973.

"Noise," by W. J. Gould, et al. ANN NY ACAD SCI 216:17-29, 1973.

"Noise," by L. Rosenhouse. NURSING CARE 7:26-28, November, 1974.

"Noise in the quiet zone." MOD HEALTHCARE, SHORT-TERM CARE ED 1:59-63, April, 1974.

"Noise! Noise! Noise!" by E. Kiester, Jr. FAMILY HEALTH 6:20-21 plus, January, 1974.

NOISE STUDIES

"Noise nuisance," by F. Holland. COUNTRY LIFE 156:630, September 5, 1974.

"Noise studies and exposure tests in metallurgy," by L. Schreiner, et al. ZENTRALBL ARBEITSMED 24:148-153, May, 1974.

"Not so silent service." HEALTH SOC SERV J 84:343, February 16, 1974.

"Planning and noise." (editorial). R SOC HEALTH J 93:58, April, 1973.

"Public health nurse and the hearing damage," by J. Jensen. SYGEPLEJERSKEN 72:6-7, July 13, 1972.

"Public health then and now. A backward glance at noise pollution," by G. Rosen. AM J PUBLIC HEALTH 64:514-517, May, 1974.

"Quiet please. Noise does affect your health and well-being," by A. Magie. LIFE HEALTH 88:14-17, March, 1973.

"Quietness as a remedy," by Z. Naumowski. PIELEG POLOZNA 4:23-24, April, 1973.

"Ralph Nader reports," by R. Nader. LADIES HOME J 91:22 plus, January, 1974.

"Sound adivce." (editorial). NURS TIMES 70:249, February 21, 1974.

"What is noise?" SCI DIGEST 75:62, May, 1974.

"What's to hear," by C. Hopkins. J ARKANSAS MED SOC 71:97-98, July, 1974.

"Why we're disturbed about noise," by C. A. Ragan, Jr. MED TIMES 102:17 plus, March, 1974.

NORTRIPTYLINE HYDROCHLORIDE

OCCUPATIONAL DEAFNESS
"Acute acoustic trauma in a locomotive engineer," by I. E. Zaslavskii.

ZH USHN NOS GORL BOLEZN 0(1):112-113, January-February, 1974.

"Benefit for hearing loss." OCCUP HEALTH 25:449-450, December, 1973.

"Further studies on industrial sudden deafness," by F. Suga. J OTO-LARYNGOL JAP 76:1373-1379, November, 1973.

"The hearing disabiltiy of the noise-damaged and the industrial injury insurance," by I. Klockhoff, et al. LAKARTIDNINGEN 71:819-822, February 27, 1974.

"Industrial noise and hearing loss," by C. O. Istre, Jr., et al. J LA STATE MED SOC 126-5-7, January, 1974.

"Intermittent noise exposure and associated damage risk to hearing of chain saw operators," by M. Schmidek, et al. AM IND HYG ASSOC J 35:152-158, March, 1974.

"Measurement of temporal shifting of hearing threshold as an evaluation of hearing loss risk under industrial conditions," by T. Malinowski, et al. OTOLARYNGOL POL 28:167-172, 1974.

"Neurosensory hypoacusia in metallurgy workers," by R. Benavides. REV MED CHIL 101:613-620, August, 1973.

"Pathogenesis of occupational hearing disorder under the combined action of general vibration and noise," by I. P. Enin. ZH USHN NOS GORL BOLEZN 0(1):53-57, January-February, 1974.

"Prevention of neurosensorial hypoacusia in metallurgic workers," by R. Benavides. REV MED CHIL 101:643-645, August, 1973.

"Should the practice for financial compensation for occupational acoustic trauma be changed?" by A. Ahlmark, et al. LAKARTIDNINGEN 70:3151-3154, September 12, 1973.

"State of auditory function in persons working since adolescence in a noisy environment," by I. B. Kramarenko, et al. VESTN OTORINO-

LARINGOL 35:93-96, January-February, 1973.

"Vestibular apparatus and occupational deafness," by P. Picart. ACTA OTORHINOLARYNGOL BELG 26:657-663, 1972.

"Welding and ear injuries," by L. Andreasson, et al. LAKARTIDNINGEN 71:2553-2554, June 19, 1974.

OCCUPATIONAL HEALTH
"The assessment of occupational noise exposure," by A. M. Martin. ANN OCCUP HYG 16(4):353-362, 1973.

"Criteria for a recommended standard—occupational exposure to noise. I. Recommendations for a noise standard," by H. M. Utidjian. J OCCUP MED 16:33-37, January, 1974.

"Danger: noise at work!" by A. S. Freese. POP MECH 142:140-145 plus, November, 1974.

"Education by example?" by R. J. Merriman. OCCUP HEALTH 26:182-183, May, 1974.

"Effect of noise exposure during primary flight training on the conventional and high frequency hearing of naval aviation officer candidates," by R. M. Robertson, et al. J ACOUST SOC AM 55:Suppl:41, 1974.

"Effect of noise on the ear following tympanoplasty in miners employed underground," by S. Stawinski. OTOLARYNGOL POL 27:751-755, 1973.

"Noise and the navy." MED TIMES 102:83, March, 1974.

"Noise at the working place," by S. Pietruck, et al. ZENTRALBL ARBEITSMED 24:139-148, May, 1974.

"Noise factors in product liability," by C. E. Wilson. QUALITY PROG 7:28-30, February, 1974.

"Noise of industrial enterprises and its effect on the population of Krivoi

Rog," by N. M. Paran'ko, et al. GIG SANIT 37:98-99, July, 1972.

"Occupational exposure to noise," by L. Hughes. J OCCUP MED 16:38, January, 1974.

"Practical problems in stopping on-the-job noise pollution," by R. D. Moran. J OCCUP MED 16:19-21, January, 1974.

"Preventing hearing loss due to excessive noise exposure," by J. Sataloff. J OCCUP MED 16:470-471, July, 1974.

"The relationship between the length of exposure to noise and the incidence of hypertension at a silo in Terran," by N. Kavoussi. MED LAV 64:292-295, July-August, 1973.

OCEANS

OFFICE NOISE
"Accoustical privacy in the landscaped office," by A. C. C. Warnock. ACOUSTICAL SOC AM J 53:1535-1543, June, 1973.

"Lightning, climate, and acoustics in large office rooms," by F. Roedler. ZENTRALBL BAKTERIOL 225:316-328, December, 1973.

"New concepts for the open office," by D. Meisner. ADM MGT 35:22-24 plus, March, 1974.

"Noise pollution in the engineering office," by R. E. Herzog. MACHINE DESIGN 45:66-71, July 26, 1973.

"Noisy trains and noisy typewriters pose different acoustical problems." ARCHIT REC 156:96-97, Mid-August, 1974.

"Should you colour your [office] sound?" by L. W. Hegvold. OPTIMUM 5,1:54-60, 1974.

OFFICE NOISE: PRINTING
"Efforts continue on reducing noise, ink, dust in pressroom," by G. B. Healey. ED & PUB 107:10, January 26, 1974.

"New way to lower pressroom noise level," by J. Hennage. ASSE J 19: 44-46, May, 1974.

"OSHA roundup: newspapers safe, noise problem mounts," by M. C. Fisk. ED & PUB 106:10 plus, December 15, 1973.

"What you ought to know and do about reducing noise in the pressroom," by W. H. Rouse. PTR/AM LITH 172:69-70 plus, April, 1974.

OTOLARYNGOLOGY

OTOLOGY

OTORHINOLARYNGOLOGY

"Otorhinolaryngology, scuba diving and hyperbaric medicine," by A. Appaix, et al. J FR OTORHINOLARYNGOL 22:559-561 plus, September, 1973.

OTOTOXICOSES

PERCEPTION

"Rhythmic structure in auditory temporal pattern perception and immediate memory," by P. T. Sturges, et al. J EXP PSYCHOL 102: 377-383, March, 1974.

"Some experiemnts relating to the perception of complex tones," by B. C. Moore. Q J EXP PSYCHOL 25:451-475, November, 1973.

PERFORMANCE
see also: Psychology

"Effect of sound on creative performance," by B. Kaltsounis. PSYCHOL REP 33:737-738, December, 1973.

"Effects of intermittent, moderate intensity noise stress on human performance," by G. C. Theologus, et al. J APP PSYCHOL 59:539-547, October, 1974.

"The effects of intermittent noise on human serial decoding performance

PERFORMANCE

"and physiological response," by D. W. Conrad. ERGONOMICS 16: 739-747, November, 1973.

"The effects of noise on the performance of simultaneous interpreters: accuracy of performance," by D. Gerver. ACTA PSYCHOL 38(3): 159-167, June, 1974.

"Performance during continuous and intermittent noise and wearing ear protection," by L. R. Hartley. J EXP PSYCHOL 102:512-516, March, 1974.

"Variables influencing performance on speech-sound discrimination tests," by A. H. Schwartz, et al. J SPEECH & HEARING RES 17: 25-32, March, 1974.

PHARMACIES

PHYSICIANS

PHYSIOLOGY
see also: Noise Research

"Action of noise on oxygen consumption in different brain structures," by S. V. Alekseev, et al. GIG SANIT 38:110-111, July, 1973.

"Assessment of acoustic trauma," (proceedings), by F. Schwetz, et al. HEFTE UNFALLHEILKD 114:197-220, 1973.

"Auditory electromyographic feedback therapy to inhibit undesired motor activity," by D. Swaan, et al. ARCH PHYS MED REHABIL 55:251-254, June, 1974.

"Auditory evoked potentials: developmental changes of threshold and amplitude following early acoustic trauma," by J. F. Willott, et al. J COMP PHYSIOL PSYCHOL 86:1-7, January, 1974.

"Cochleo vestibular disturbances in vibration disease," by M. A. Nekhorosheva, et al. GIG TR PROF ZABOL 17(7):9-11, 1973.

"Comparative characteristics of the action of moderate levels of constant and interrupted noise on certain body functions," by L. A. Oleshkevich. GIG SANIT 38:95-97, August, 1973.

"Conditioning treatment of enuresis: auditory intensity," by G. C. Young, et al. BEHAV RES THER 11:411-416, November, 1973.

"A contribution to the physiology of the perilymph. 3. On the origin of noise-induced hearing loss," by E. A. Schnieder. ANN OTOL RHINOL LARYNGOL 83:406-412, May-June, 1974.

"Damage due to noise," by A. Delmas. BULL ACAD NATL MED 157 (4):272-276, 1973.

"Diagnosis of post-trauma vertigo," by A. Montandon. J FR OTORHINOLARYNGOL 22:647-649, September, 1973.

"The diagnostic contribution of EEG in the assessment of noise induced hearing losses," by H. G. Dieroff, et al. Z LARYNGOL RHINOL OTOL 52:908-914, December, 1973.

"Echographic diagnosis of the ciliary body detachment," by B. N. Alekseev. VESTN OFTALMOL 4:20-27, 1973.

"Effect of infrasonics on the body," by E. N. Malyshev, et al. GIG SANIT 39:27-30, March, 1974.

"Effect of lithium carbonate and alpha-methyl-p-tyrosine on audiogenic seizure intensity," (letter), by P. C. Jobe, et al. J PHARM PHARMACOL 25:830-831, October, 1973.

"Effect of noise and ototoxic substances on previously damaged ears," (proceedings), by M. Quante. ARCH KLIN EXP OHREN NASEN KEHLKOPFHEILKD 205:266-269, December 17, 1973.

"Effect of noise on the general immunological reactivity of the body," by M. L. Khaimovich. GIG SANIT 38:96-98, February, 1973.

"The effect of noise on human sensations," by L. A. Oleshkevich. VRACH DELO 3:127-130, 1973.

PHYSIOLOGY

"Effect of noise on information processing processes and on simple motor reaction," by B. Stefenov, et al. GIG SANIT 37:84-86, September, 1972.

"The effects of intermittent noise on human serial decoding performance and physiological responses," by D. W. Conrad ERGONOMICS 16:739-747, November, 1973.

"Etiological factors in hearing loss in tangential missile wounds of the head," by A. Adeloye, et al. LARYNGOSCOPE 84:126-131, January, 1974.

"Health hazard: sound pollution." MUS J 31:27 plus, December, 1973.

"Hearing loss among Baffin Zone Eskimos—a preliminary report," by J. D. Baxter, et al. CAN J OTOLARYNGOL 1:337-343, 1972.

"Hearing loss: ways to avoid it, or live with it; interview," by R. E. Jordan. U.S. NEWS 76:48-50, January 21, 1974.

"Inhibition of fetal osteogenesis by maternal noise stress," by W. F. Geber. FED PROC 32:2101-2104, November, 1973.

"Medical treatment of deafness," by J. P. Secretan. REV MED SUISSE ROMANDE 93:975-977, December, 1973.

"Monaural and binaural speech perception through hearing aids under noise and reverberation with normal and hearing-impaired listeners," by A. K. Nabelek, et al. J SPEECH & HEARING RES 17:724-730, December, 1974.

"Morphological changes in the cochlea in experimental noise trauma: phase contrast microscopy," by E. Sh. Suladze, et al. VESTN OTORINOLARINGOL 35:23-26, May-June, 1973.

"New method of diagnosis of joint diseases—arthrophonography," by M. A. Iasinovskii, et al. KLIN MED 51:25-28, July, 1973.

"Noise-induced hearing loss," by F. J. Dittrich. J AM OSTEOPATH ASSOC 73:446-449, February, 1974.

PHYSIOLOGY

"Noise-induced hearing loss and presbyacusis," by J. H. Macrae. AUDIOLOGY 10:323-333, September-December, 1971.

"Physiological correlates of auditory stimulus periodicity," by J. E. Hind. AUDIOLOGY 11:42-57, January-April, 1972.

"Physiological effects of intermittent noise," (proceedings), by Y. Osada, et al. J PHYSIOL SOC JAP 35:460, August-September, 1973.

"Physiological effects of noise. An overview," by B. L. Welch. FED PROC 32:2091-2092, November, 1973.

"Problems arising from studies concerning the convergence of noise-related hardness of hearing and hearing problems of other origins with special attention paid to agerelated hardness of hearing," by T. Brusis. Z LARYNGOL RHINOL OTOL 52:915-929, December, 1973.

"Pure tone audiometric picture of noise induced deafness," by T. Brusis. Z LARYNGOL RHINOL OTOL 52:673-680, September, 1973.

"Solution of the noise problem with the Aue II artificial kidney," by J. Iversen, et al. Z UROL NEPHROL 66:919-920, December, 1973.

"Some aspects of the problem of adaptation to noise," by E. Ts. Andreeva-Galanina, et al. GIG SANIT 38:34-37, December, 1973.

"Sound intensity and good health," by H. Haggerty. PHYS TEACH 12:421-423, October, 1974.

"Sound wave receiving mechanism seen from the aspect of comparative physiology," by Y. Katsuki. J OTOLARYNGOL JAP 76:1297-1300, October, 1973.

"Spectrographic analysis of fundamental frequency and hoarseness before and after vocal rehabilitation," by M. Cooper. J SPEECH HEAR DISORD 39(3):286-297, August, 1974.

"Stimulus intensity and recency contrasts and orienting response strength," by D. C. Edwards. PSYCHOPHYSIOLOGY 11(5):543-

PHYSIOLOGY

547, September, 1974.

"Stria ultrastructure and vessel transport in acoustic trauma," by A. J. Duvall, III, et al. ANN OTOL RHINOL LARYNGOL 83:498-514, July-August, 1974.

"Study of cerebral circulation under the separate and joint action of intensive noise and physical load," by I. B. Evdokimova, et al. GIG TR PROF ZABOL 17:1-5, July, 1973.

"Therapy of patients with acoustic trauma," by G. A. Dokukina, et al. VOEN MED ZH 8:35-38, August, 1973.

"Ultrastructure of the spiral organ of the cochlea under normal conditions and following an experimental acoustic trauma," by O. Sh. Goniashvili. VESTN OTORINOLARINGOL 35:58-63, September-October, 1973.

"The validity of the 'energy principle' for noise-induced hearing loss," by H. Scheiblechner. AUDIOLOGY 13:93-111, March-April, 1974.

"The value of directional audiometry in the assessment of the central components in noise induced hearing loss," by H. G. Dieroff. Z LARYNGOL RHINOL OTOL 52:681-686, September, 1973.

PIPELINES

"Vibration and noise in piping systems," by S. J. Shuey, et al. POWER ENG 77:42-44, June, 1973.

PLANTS

"Controlling in-plant noise," by K. M. Hankel. AUTOMATION 21:86-90, April, 1974.

"Downing the plant's din." CHEM ENG 80:30 plus, December 24, 1973.

"Elastomer sprag cuts plant noise." PURCHASING 76-57, January 8, 1974.

"Interdisciplinary plant-noise control," by A. Thumann. CHEM ENG 81:120 plus, August 19, 1974.

PLANTS

"Large gas handling plants in noise control," by T. Dear. CHEM ENG PROG 70:65-68, February, 1974.

"Noise-control design for process plants," by S. C. Lou. CHEM ENG 80: 77-82, November 26, 1973.

"Noise control in coal preparation plants," by G. Petersen. MIN CONG J 60:30-36, January, 1974.

"Noise factor in the main metallurgical production shops and measures for its control," by L. A. Sobolevskaia. GIG TR PROF ZABOL 16: 3-7, July, 1972.

"Predict plant noise problems," by R. S. Norman. HYDROCARBON PROCESS 52:89-91, October, 1973.

"Reducing noise in food plants," by R. K. Miller. FOOD ENG 46:75-76, February, 1974.

"Solving typical plant noise problems." PLANT ENG 27:54 plus, December 13, 1973.

"Sound-technical viewpoints for the planning and realization of heating centers," by H. Schmitz. GESUND ING 92:141-146, May, 1971.

"Stop plant noise at the source or along the way," by C. L. Meteer. AUTOMATION 21:58-61, July, 1974.

PLANTS: AUTOMOBILE
"GM's noise program: broad, wide and deep," by C. A. Gottesman. AUTOMOTIVE IND 151:54 plus, November 1, 1974.

PLANTS: CELLULOSE-PAPER
"New valve-silencer reduces noise level on mill steam line [Ontario-Minnesota pulp & paper]." PULP & PA 48:122, May, 1974.

PLANTS: CHEMICAL
"Carbide plant too noisy?" CHEM W 114:15, March 13, 1974.

"Personal noise dosimetry in refinery and chemical plants," by A. H.

PLANTS: CHEMICAL

Diserens. J OCCUP MED 16:255-257, April, 1974.

PLANTS: CLEANING

PNEUMATIC MACHINERY AND TOOLS
see: Compressed Air Noise

PRESSURE BOXES

PROTECTION
"Antinoise ear plug made of film porolon," by V. Ia. Gapanovich, et al. GIG TR PROF ZABOL 16:54, July, 1972.

"Attenuation characteristics of recreational helmets," by F. H. Bess, et al. ANN OTOL RHINOL LARYNGOL 83:119-124, January-February, 1974.

"Considerations on the tolerance of workers for ear protective devices," by J. P. Pepersack. ARCH BELG MED SOC 31:179-183, March, 1973.

"Controlling industrial noise: administrative controls and hearing protection," by C. H. Wiek. MANUF ENG & MGT 71:32-35, July, 1973.

"Effectiveness of different ear protectors in protecting the employee from over exposure in industrial environments," by J. E. Stephenson, et al. J ACOUST SOC AM 54(1):301, 1973.

"High frequency attenuation characteristics of ear protectors," by F. H. Bess, et al. J ACOUST SOC AM 54(1):328, 1973.

"High frequency attenuation characteristics of ear protectors," by T. H. Townsend, et al. J OCCUP MED 15:888-891, November, 1973.

"How R. J. Reynolds protects workers' hearing." MGT R 62:64-66, July, 1973.

"Individual hearing protection—survey of Berlin's noisy factories," by P. Moch. ZENTRALBL ARBEITSMED 23:33-38, February, 1973.

PROTECTION

"Noise exposure risks lessened; CEL Noise Dosimeter." ENGINEER 237: 23, July 12, 1973.

"Noise: the government's view," by R. E. Train. MED TIMES 102:63-64, March, 1974.

"A note on the protection afforded by hearing protectors—implications of the energy principle," by D. Else. ANN OCCUP HYG 16:81-83, April, 1973.

"Performance during continuous and intermittent noise and wearing ear protection," by L. R. Hartley. J EXP SPYCHOL 102:512-516, March, 1974.

"Studies on the perception of acoustic signals under noise-protection conditions in track-packing work, and corresponding conclusions," by K. Jungsbluth, et al. ZENTRALBL ARBEITSMED 24:153-156, May, 1974.

"Workers' choice . . . ear defenders." OCCUP HEALTH 26:474, December, 1974.

PSILOCYBINE

PSYCHOLOGY

"Aircraft noise and psychiatric morbidity," by F. Gattoni, et al. PSYCHOL MED 3:516-520, November, 1973.

"Arousal and recall: effects of noise on two retrieval strategies," by S. Schwartz. J EXP PSYCHOL 102:896-898, May, 1974.

"Comprehensive clinical and psychological studies of patients exposed to chronic acoustic trauma," by S. Klonowski, et al. POL TYG LEK 29:313-315, February 25, 1974.

"Controlling technically produced noise to reduce psychological stress," by G. Carlestam. IMPACT OF SCIENCE ON SOCIETY 23,3:237-248, July-September, 1973.

"Effect of noise on intellectual performance," by N. D. Weinstein.

J APP PSYCHOL 59:548-554, October, 1974.

"Effect of noise on the Stroop Test," by L. R. Hartley, et al. J EXP PSYCHOL 102:62-66, January, 1974.

"Effect of white noise on the reaction time of mentally retarded subjects," by C. M. Mlezejeski. AM J MEN DEFICIENCY 79:39-43, July, 1974.

"Effects of airplane noise on health: an examination of three hypotheses," by D. B. Graeven. HEALTH & SOC BEHAV 15:336-343, December, 1974.

"The effects of intelligence quotient and extraneous stimulation on incidental learning," by R. Forehand, et al. J MENT DEFIC RES 17:24-27, March, 1973.

"Effects of intermittent, moderate intensity noise stress on human performance," by G. C. Theologus, et al. J APP PSYCHOL 59:539-547, October, 1974.

"Effects of white noise on the frequency of stuttering," by S. F. Garber, et al. J SPEECH & HEARING RES 17:73-79, March, 1974.

"Generalization of stimulus control in a summer camp," by G. E. Taylor, Jr., et al. PSYCHOL REPT 34:419-423, April, 1974.

"Helplessness, stress level, and the coronary-prone behavior pattern," by D. S. Krantz, et al. J EXP SOC PSYCHOL 10:284-300, May, 1974.

"Psychological effects of exposure to high industrial noise: a field study," by E. Gulian. J ACOUST SOC AM 55:Suppl:68, 1974.

"Some effects of noise on the speaking behavior of stutterers," by E. G. Conture. J SPEECH & HEARING RES 17:714-723, December, 1974.

"Some implications regarding high frequency hearing loss in school-age children," by R. L. Cozad, et al. J SCH HEALTH 44:92-96, February, 1974.

PSYCHOLOGY

"Variables influencing performance on speech-sound discrimination tests," by A. H. Schwartz, et al. J SPEECH & HEARING RES 17: 25-32, March, 1974.

PSYCHOTICS
"Environmental noise level as a factor in the treatment of hospitalized schizophrenics," by M. F. Ozerengin. DIS NERV SYST 35(5):241-245, 1974.

RADIO NOISE
"Radio noise of terrestrial origin; abstracts of papers." RADIO SCI 8: 613-621, June, 1973.

RAILWAY NOISE
"EPA proposes rail noise standards." AIR POLLUTION CONTROL ASSN J 24:881, September, 1974.

"Evaluation of the decrease in auditory function in railroad workers according to tonal audiographic data," by I. E. Zaslavskii. GIG TR PROF ZABOL 16:44-47, July, 1972.

"A methodology for assessing potential community impact resulting from noise emitted by railroad yard operations," by J. W. Swing, et al. J ACOUST SOC AM 55(2):465-466, 1974.

"Noise and vibration of resiliently supported track slabs," by E. K. Bender. ACOUSTICAL SOC AM J 55:259-268, February, 1974.

"Noisy trains and noisy typewriters pose different acoustical problems." ARCHIT REC 156:96-97, Mid-August, 1974.

"Rubber pad quiets wheel squeal." MACHINE DESIGN 46:42, September 5, 1974.

RAINDROPS

RECREATIONAL
"Recreational noise: implications for potential hearing loss to participants," by J. H. Shirreffs. J SCH HEALTH 44:548-550, December, 1974.

RECREATIONAL

"Sports, noisy and quiet," (letter), by R. H. Nuenke. N ENGL J MED 290:523, February 28, 1974.

RESPIRATORY SYSTEM
see: Noise Research

SCHOOLS

"The acoustics of educational facilities," by E. P. Caffarella. AUDIO-VISUAL INSTRUCTION 18,10:10-11, December, 1973.

"Jet noise at schools near Los Angeles International Airport," by S. R. Lane, et al. J ACOUST SOC AM 56:127-131, July, 1974.

"Noise environment of a typical school classroom due to the operation of utility helicopters," by D. A. Hilton, et al. J ACOUST SOC AM 55:Suppl:37, 1974.

"Unobtrusive-sound reinforcement for an open-plan school." ARCHIT REC 156:151-152, September, 1974.

SLEEP

"Arousal from sleep: the differential effect of frequencies equated for loudness," by T. E. Levere, et al. PHYSIOL BEHAV 12:573-582, April, 1974.

"Body movements in sleep during 30-day exposure to tone pulse," by A. G. Muzet, et al. PSYCHOPHYSIOLOGY 11:27-34, January, 1974.

"The effect of white noise on the somatosensory evoked response in sleeping newborn infants," by P. H. Wolff, et al. ELECTROEN-CEPHALOGR CLIN NEUROPHYSIOL 37:269-274, September, 1974.

"The recovery cycle of the averaged auditory evoked response during sleep in normal children," by E. M. Ornitz, et al. ELECTROEN-CEPHALOGR CLIN NEUROPHYSIOL 37:113-122, August, 1974.

SMALL BOREARMS

SNOWMOBILES

"Noise-induced hearing loss and snowmobiles," by F. H. Bess, et al. ARCH OTOLARYNGOL 99:45-51, January, 1974.

"Relation of noise measurements to temporary threshold shift in snowmobile users," by R. B. Chaney, Jr., et al. ACOUSTICAL SOC AM J 54:1219-1223, November, 1973.

"Silencing the snowmobile." AUTOMOTIVE ENG 81:72-73 plus, September, 1973.

SOCIOLOGY
see: Behavior
Physiology
Psychology

SODIUM POTASSIUM

SODIUM SALICYLATE

SONIC BOOM
"Photographic analysis of human startle reaction to sonic booms," by R. I. Thackray, et al. AEROSP MED 45:803-806, August, 1974.

SOUNDPROOFING
"ABC's of sound reinforcement," by M. Koller. RADIO-ELECTR 45: 40, August, 1974.

"Acoustical building panels quieten turbine compressor stations." PIPELINE & GAS J 201:42 plus, May, 1974.

"Acoustical floor covering comes of age." AM SCH & UNIV 46,3:35-36, November, 1973.

"Acoustical value of carpeting," by D. B. Parlin. BUILDING OPER MANAGE 21:48 plus, September, 1974.

"Barriers for noise control," by J. N. Macduff. MECH ENG 96:26-31, August, 1974.

SOUNDPROOFING

"Big noises are being heard as industry considers the cost," by C. Beaton. ENGINEER 238-38-39, February 7, 1974.

"Compressors beat new noise-law levels." ELEC WORLD 180:85, July 15, 1973.

"Controlling industrial noise; acoustic materials and enclosures," by C. H. Wick. MANUF ENG & MGT 70:30-33, June, 1973.

"Controlling noisy washer-dryer systems." PLANT ENG 28:48, February 7, 1974.

"Curbing noise with partial enclosures," by W. G. Phillips, et al. MACHINE DESIGN 46:107-110, April 4, 1974.

"Designing safety into underground mining equipment; noise abatement," by C. Holvenstot. MIN CONG J 59:39-43, September, 1973.

"Do not disturb [O'Hare international tower hotel]." ENGIN N 190:12, May 24, 1973.

"Four ways materials combat noise pollution," by K. H. Miska. MATERIALS ENG 79:20-23, June, 1974.

"Get the lead in, get noise out [sound-insulated enclosures]." FACTORY 6:11, November, 1973.

"Highway noise and acoustical buffer zones," by A. Zulfacar, et al. AM SOC C E PROC 100:389-401, May, 1974.

"Holding down the decibel count." PLASTICS ENG 30:23, April, 1974.

"If you can't beat noise, baffle it." ENGINEER 237:23, September 13, 1973.

"Kinematic sound screen; unique solution to highway noise abatement," by J. B. Hauskins, Jr. AM SOC C E PROC 100:169-178, February, 1974.

"Machine noise is reduced by fitting sliding shutters." ENGINEER 238:

SOUNDPROOFING

23, June 6, 1974.

"Materials that build a box around noise," by B. D. Wakefield. IRON AGE 214:53-54 plus, July 15, 1974.

"New concepts for the open office," by D. Meisner. ADM MGT 35:22-24 plus, March, 1974.

"New materials reduce noise, vibration." AUTOMOTIVE ENG 81:10, March, 1973.

"New way to lower pressroom noise level," by J. Hennage. ASSE J 19:44-46, May, 1974.

"Noise control; a common-sense approach," by R. L. Lowery. MECH ENG 95:26-31, June, 1973.

"Noise control does not have to be a problem: Moduline system." ENGINEER 239:22, July 4, 1974.

"Noise control enclosure improves dryer efficiency." ROADS & STS 116:142 plus, September, 1973.

"Noise control in coal preparation plants," by G. Petersen. MIN CONG J 60:30-36, January, 1974.

"Noise-reducing punch-press guard," by R. S. Florczyk. PLANT ENG 27:158-159, October 18, 1973.

"Noise reduction by barriers," by U. J. Kurze. ACOUSTICAL SOC AM J 55:504-518, March, 1974.

"Noisy trains and noisy typewriters pose different acoustical problems: so get different acoustical treatment." ARCHIT REC 156:96-97, Mid-August, 1974.

"Open-cell urethane foams offer pluses for noise control." PRODUCT ENG 45:37-38, April, 1974.

"Reducing noise in food plants," by R. K. Miller. FOOD ENG 46:75-76,

SOUNDPROOFING

February, 1974.

"Shut up those sudden noises." ENGINEER 237:23, November 29, 1973.

"Silencing of generators." ENGINEER 238:18, February 28, 1974.

"Silent cabins, silencers, machine enclosures." ENGINEER 239:18, September 12, 1974.

"Soundproofing your engine compartment," by J. Martenhoff. MOTOR B & S 134:67-69, July, 1974.

SPEECH

"Methods of acoustical analysis of speech," by M. Wajskop. ACTA OTO-RHINOLARYNGOL BELG 26:741-756, 1972.

"Monaural and binaural speech perception through hearing aids under noise reverberation with normal and hearing-impaired listeners," by A. K. Nabelek, et al. J SPEECH & HEARING RES 17:724-739, December, 1974.

"Presbyacusis. VI. Masking of speech," by K. Jokinen. ACTA OTO-LARYNGOL 76:426-430, December, 1973.

"Recent physical examination technics of speech," (proceedings), by F. J. Landwehr, et al. ARCH KLIN EXP OHREN NASEN KEHLKOPF-HEILKD 205:388-391, December 17, 1973.

SST
see: Sonic Boom

STATISTICS

"Statistical analysis of continuous data records," by R. B. Corotis. AM SOC C E PROC 100:195-206, February, 1974.

"Statistical analysis of telephone noise," by B. W. Stuck, et al. BELL SYSTEM TECH J 53:1263-1320, September, 1974.

SUBWAYS

"Rubber pad quiets wheel squeal." MACHINE DESIGN 46:42, September 5, 1974.

"Statistical analysis of continuous data records," by R. B. Corotis. AM SOC C E PROC 100:195-206, February, 1974.

SURGERY
see: Acoustic Nerve

SYMPOSIA

"Conference on vehicle noise and the designer, Hatfield, England." ENGINEER 238:27, May 2, 1974.

"Noise abatement, problem and progress, symposium." DIESEL EQUIP SUPT 52:42-44, June, 1974.

"Noise pollution seminar." MECH ENG 96:42, January, 1974.

"Symposium: new data for noise standards. I. New data for noise standards," by D. Henderson, et al. LARYNGOSCOPE 84:714-721, May, 1974.

"Symposium: new data for noise standards. IV. The physiological effects of priming for audiogenic seizures in mice," by J. C. Saunders. LARYNGOSCOPE 84:750-758, May, 1974.

TELEPHONES

"Statistical analysis of telephone noise," by B. W. Stuck, et al. BELL SYSTEM TECH J 53:1263-1320, September, 1974.

TEMPORARY THRESHOLD SHIFT

"Decay of temporary threshold shift in noise; monaural chinchillas," by J. H. Mills, et al. J SPEECH & HEARING RES 16:267-270, June, 1973.

"Effects of varying levels of interruption of temporary threshold shift," by M. E. Schmidek. J ACOUST SOC AM 55:Suppl:40, 1974.

"Human temporary threshold shift from 16-hour noise exposures," by W.

Meinick. ARCH OTOLARYNGOL 100:180-189, September, 1974.

"Measurement of temporal shifting of hearing threshold as an evaluation of hearing loss risk under industrial conditions," by T. Malinowski, et al. OTOLARYNGOL POL 28:167-172, 1974.

"Noise-induced threshold shift in the parakeet (Melopsittacus undulatus)," by J. Saunders, et al. PROC NATL ACAD SCI USA 71: 1962-1965, May, 1974.

"The relation between temporary threshold shift and permanent threshold shift in rhesus monkeys exposed to impulse noise," by G. A. Luz, et al. ACTA OTOLARYNGOL [Suppl] 1-15, 1973.

"Relation of noise measurements to temporary threshold shift in snowmobile users," by R. B. Chaney, Jr., et al. J ACOUST SOC AM 54: 1219-1223, November, 1973.

"Temporary threshold shift from a toy cap gun; Bekesy technique," by L. Marshall, et al. J SPEECH & HEARING DIS 39:163-168, May, 1974.

"Threshold shifts produced by exposure to noise in chinchillas with noise-induced hearing losses," by J. H. Mills. J SPEECH HEAR RES 16:700-708, December, 1973.

THUNDER
see: Lightning

TINNITUS

TRACTOR OPERATORS

TRACTORS
"Health evaluation of the noise of automotive agricultural machines," by V. V. Vlasenko. GIG TR PROF ZABOL 16:52-54, July, 1972.

"Noise and vibrations of agriculture tractors and measures of prevention," by D. Petrovic. NAR ZDRAV 29:110-115, April, 1973.

TRAFFIC NOISE

"Bituminous overlay reduces traffic noise," by H. O. Klossner. PUB WORKS 104:82-83, July, 1973.

"Effect of an interstate highway on urban area noise levels," by J. E. Heer, Jr., et al. PUB WORKS 105:54-58, January, 1974.

"Effects of a traffic noise background on judgements of aircraft noise," by C. A. Powell. J ACOUST SOC AM 55:Suppl:68, 1974.

"Environmental noise classifier and its use," by J. Donovan. AUDIO ENG SOC J 22:528-532, September, 1974.

"First plug your ears." ECONOMIST 248:49-50, September 1, 1973.

"Highway noise and acoustical buffer zones," by A. Zulfacar, et al. AM SOC C E PROC 100:389-401, May, 1974.

"Increased noise as an element of compensation in condemnation proceedings," by W. R. Theiss. APPRAISAL J 42:134-138, January, 1974.

"Kinematic sound screen; unique solution to highway noise abatement," by J. B. Hauskins, Jr. AM SOC C E PROC 100:169-178, February, 1974.

"Noise and the highway patrolman," by W. R. Pierson, et al. J OCCUP MED 15:892-893, November, 1973.

"The problem of traffic noise," by D. J. Fisk. R SOC HEALTH J 93:289-290 plus, December, 1973.

"Surface transportation noise and its control," by J. E. Wesler. AIR POLLUTION CONTROL ASSN J 23:701-703, August, 1973.

"Theoretical prediction of highway noise fluctuations," by A. H. Marcus. J ACOUST SOC AM 56:132-136, July, 1974.

"Traffic noise and overheating in offices." BUILD RES ESTAB DIGEST 162:1-4, February, 1974.

TRANSMISSIONS

TRANSMISSIONS

TRIVASTAL

TURBINES

TURBINES: GAS

"Aircraft environmental problems," by V. L. Blumenthal, et al. J AIRCRAFT 10:529-537, September, 1973.

"Business flying faces new challenges; Hawker Siddeley, Rolls to test HS.125 with sound suppression," by H. J. Coleman. AVIATION W 99:57-58, September 24, 1973.

"Designing small gas turbine engines for low noise and clean exhaust," by H. C. Eatock, et al. J AIRCRAFT 11:616-622, October, 1974.

"Development of sonic inlets for turbofan engines," by F. Klujber. J AIRCRAFT 10:579-586, October, 1973.

"Diffraction of a plane wave by a half plane in a subsonic and supersonic medium," by S. M. Candel. ACOUSTICAL SOC AM J 54:1008-1016, October, 1973.

"Effect of ejector spacing an ejector-jet noise characteristics," by D. Tirumalesa. ACOUSTICAL SOC AM J 56:911-916, September, 1974.

"Geared fan engine systems; their advantages and potential reliability," by T. A. Lyon, et al. J AIRCRAFT 10:361-365, June, 1973.

"Jet engine exhaust noise due to rough combustion and nonsteady aerodynamic sources," by E. G. Plett, et al. ACOUSTICAL SOC AM J 56:516-522, August, 1974.

"NASA JT8D refan program nears end," by M. L. Yaffee. AVIATION W 101:46-47, July 22, 1974.

"Owl's wing a noise inhibitor," by P. Soderman. MECH ENG 95:56, October, 1973.

TURBINES: GAS

"Test and evaluation of a quiet helicopter configuration HH-43B," by M. A. Bowes. ACOUSTICAL SOC AM J 54:1214-1218, November, 1973.

"Unified analysis of fan stator noise," by D. B. Hanson. ACOUSTICAL SOC AM J 54:1571-1591, December, 1973.

URBAN NOISE

"Automatic urban noise monitoring and analysis system," by J. E. K. Foreman, et al. ACOUSTICAL SOC AM J 55:1358-1359, June, 1974.

"Comparison of inside and outside noise measurements in various urban environments," by D. E. Bishop. J ACOUST SOC AM 55(2):465, 1974.

"Data on the hygienic evaluation of city noise," by S. A. Soldatkina, et al. GIG SANIT 38:16-20, March, 1973.

"Effect of an interstate highway on urban area noise levels," by J. E. Heer, Jr., et al. PUB WORKS 105:54-58, January, 1974.

"Helicopter noise experiments in an urban environment," by W. A. Kinney, et al. ACOUSTICAL SOC AM J 56:332-337, August, 1974.

"Man-made noise in urban environments and transportation systems; models and measurement," by D. Middleton. IEEE TRANS COM 21:1232-1241, November, 1973.

"Measuring procedure for urban noise in the center of Mexico City," by F. Groenwold, et al. J ACOUST SOC AM 55(2):465, 1974.

"Method of composing a noise map of a city," by O. S. Rastorguev, et al. GIG SANIT 37:62-65, October, 1972.

"Noise measurement in planning, by H. Williams. Noise from urban roads, by E. Rowlands. Traffic noise: some practical implications, by W. Allen, et al." ROY TOWN PLAN INST J 59:7-16, January, 1973.

"Noise propagation in cellular urban and industrial spaces," by H. G.

URBAN NOISE

Davies, et al. ACOUSTICAL SOC AM J 54:1565-1570, December, 1973.

"Reverberation in a city street," by D. Aylor, et al. ACOUSTICAL SOC AM J 54:1754-1755, December, 1973.

"Role of multiple reflections and reverberation in urban noise propagation," by R. H. Lyon. ACOUSTICAL SOC AM J 55:493-503, March, 1974.

"Studies on noise in small cities. 2. City noise in Karatsu City," by S. Sato, et al. JAP J HYG 28:425-428, October, 1973.

--3. Community reaction to noise in Karatsu City," by H. Miura, et al. JAP J HYG 28:429-436, October, 1973.

"Urban environment: noise and transportation. Environmental backlash—the urban paradox, noise and transportation. A. G. Greenwald; Regulation—local, state and federal. J. V. Tunner; Standards and controls. A. F. Meyer, Jr.; Compliance and technology. W. J. K. Gibson; Rights, remedies and planning. D. C. McGrath, Jr." NATURAL RESOURCES LAW 7:293-323, Spring, 1974.

URBAN PLANNING
see also: Urban Noise

"Participation in urban planning: the Barnsbury case [book review]," by R. Mordey. ROY TOWN PLAN INST J 59:2-3, January, 1974.

"Planning for airports in urban environments—a survey of the problem and its possible solutions," by M. L. Dworkin. URBAN LAW 5:472-503, Summer, 1973.

VALVE NOISE
"Control valve muffles high-noise operations." CHEM ENG 80:56, August 6, 1973.

"Control-valve noise yields to research," by R. Nugent. POWER 117:69-71, July, 1973.

VALVE NOISE

"Guide to control valve noise; with chart," by J. A. Dillon. INSTR & CONTROL SYSTEMS 47:95-104, September, 1974.

"Maze silences valve noise." MACHINE DESIGN 46:38, September 19, 1974.

"New valve-silencer reduces noise level on mill steam line [Ontario-Minnesota pulp & paper]." PULP & PA 48:122, May, 1974.

VIBRATION

"Aircraft noise induced vibration in fifteen residences near Seattle-Tacoma International Airport," by S. M. Cant, et al. AM IND HYG ASSOC J 34:463-468, October, 1973.

"Biological modeling and criteria for standardization of whole-body vibration and noise," by E. Ts. Andreeva-Galanina, et al. VESTN AKAD MED NAUK SSSR 28:30-37, 1973.

"Controlled vibration feeds dry mixes, eliminates noise problem [Modern maid food products]." FOOD PROCESSING 35:93, May, 1974.

"Industrial noise and vibration in sewing industry enterprises and an evaluation of measures to decrease them," by V. F. Rudenko, et al. GIG TR PROF ZABOL 17:36-38, July, 1973.

"Mechanical equipment noise and vibration control," by L. F. Yerges. HEATING PIPING 45:61-66, July, 1973.

"New materials reduce noise, vibration." AUTOMOTIVE ENG 81:10, March, 1973.

"Noise and vibration analysis of an impact forming machine," by A. E. M. Osman, et al. J ENG IND 96:233-240, February, 1974.

"Noise and vibration hazards." JAP J HYG 29:162-169, April, 1974.

"Noise and vibration of resiliently supported track slabs," by E. K. Bender. ACOUSTICAL SOC AM J 55:259-268, February, 1974.

"Noise and vibrations of agriculture tractors and measures of preven-

VIBRATION

tion," by D. Petrovic. NAR ZDRAV 29:110-115, April, 1973.

"Noise control versus shock and vibration engineering," by C. T. Morrow. ACOUSTICAL SOC AM J 55:695-699, April, 1974.

"Pathogenesis of occupational hearing disorder under the combined action of general vibration and noise," by I. P. Enin. ZH USHN NOS GORL BOLEZN 0(1):53-57, January-February, 1974.

"Vibration and noise in piping systems," by S. J. Shuey, et al. POWER ENG 77:42-44, June, 1973.

"Vibrations during construction operations," by J. F. Wiss. AM SOC C E PROC 100:239-246, September, 1974.

VISION

"Diagnostic significance of acoustic measurement of the eyes in unilateral exophthalmos," by G. V. Kruzhkova. VESTN OFTALMOL 2:88-89, March-April, 1974.

"The effect of noise on visual fields," by J. E. Letourneau, et al. EYE EAR NOSE THROAT MON 53:49-51, February, 1974.

WEAVERS

WHITE NOISE

"Effect of white noise on the reaction time of mentally retarded subjects," by C. M. Miezejeski. AM J MEN DEFICIENCY 79:39-43, July, 1974.

"The effect of white noise on the somatosensory evoked response in sleeping newborn infants," by P. H. Wolff, et al. ELECTROEN-CEPHALOGR CLIN NEUROPHYSIOL 37:269-274, September, 1974.

"Effects of white noise on the frequency of stuttering," by S. F. Garber, et al. J SPEECH & HEARING RES 17:73-79, March, 1974.

"The effects of white noise on PB scores of normal and hearing-impaired listeners," by R. W. Keith, et al. AUDIOLOGY 11:177-186, May-

WHITE NOISE

August, 1972.

"Shock tolerance in rats as a function of white noise," by M. Cunningham, et al. PSYCHOL REP 34:711-713, June, 1974.

"Spatial stimulus generalization as a function of white noise and activation level," by R. E. Thayer, et al. J EXP PSYCHOL 102:539-542, March, 1974.

WIND

"How wind noise affects human hearing." MACHINE DESIGN 45:6, November 1, 1973.

WORKSHOPS

YOUTH

"Characteristics of the cardiovascular system of adolescent workers subjected to the action of stable industrial noise," by E. A. Gel'tishcheva. GIG TR PROF ZABOL 16:29-33, July, 1973.

"Effect of different parameters of industrial noise on the auditory analyzer and the central nervous system of adolescent workers," by E. A. Gel'tishcheva. GIG TR PROF ZABOL 17:5-9, July, 1973.

"Marked acoustical signs of voice virilization in girls," by A. Pruszewicz, et al. FOLIA PHONIATR 25:331-341, 1973.

"Temporary hearing losses in teenagers attending repeated rock-and-roll sessions," by R. F. Ulrich. ACTA OTO-LARYNGOL 77(1-2):51-55, 1974.

AUTHOR INDEX

Adeloye, A. 26
Again, T. C. 43
Agnetti, V. 49
Ahlmark, A. 57, 69
Aigner, A. 60
Alberti, P. W. 47
Alekseev, B. N. 20
Alekseev, S. V. 6, 12, 21
Alexandre, A. 1, 58
Allen, D. S. 59
Allen, J. 5
Anderson, D. J. 54
Andreasson, L. 68
Andreeva-Galanina, E. Ts. 11, 58
Anichin, V. F. 19
Anthony, A. 10
Appaix, A. 49
Aran, J. M. 29
Arseni, C. 12
Atherley, G. R. C. 16, 45
Atweh, S. F. 51
Austin, D. W. 46
Aylesworth, T. G. 1
Aylor, D. 56

Bailey, J. R. 30
Balabanov, O. V. 39
Bamberger-Bozo, C. 35
Baron, G. 19
Bartholomew, R. 1
Baxter, J. D. 30
Beatson, C. 11
Beck, G. 1
Benavides, R. 41, 52, 53

Bench, J. 61
Bender, E. K. 43
Berkovitch, I. 54
Berkovitz, R. 19
Bess, F. H. 9, 31, 45
Biagoveshchensiaia, N. S. 66
Billewicz-Stankiewicz, J. 22, 23
Bishop, D. E. 14
Blennow, G. 46
Blumethal, V. L. 7
Board, G. 66
Bock, G. R. 52
Bohme, H. R. 9
Boole, R. A. 64
Booth, R. T. 37
Borchardt, U. 63
Borsky, P. N. 8
Bowes, M. A. 64
Braden, M. 9
Bradley, W. A. 69
Branch, M. C. 40
Brewer, W. E. 24
Briffa, F. E. J. 14
Brines, G. L. 53
Brugge, J. F. 56
Brusis, T. 53
Bryson, F. E. 64
Bulban, E. J. 7
Burd, A. 30
Burgess, J. C. 67
Burns, W. 1
Butler, R. A. 19

Caccavari, C. 10

Caffarella, Jr., E. P. 6
Callaway, V. E. 55
Candel, S. M. 19
Cant, S. M. 7
Caplan, F. 12, 48
Carlson, J. P. 26
Carruth, H. 27
Castaneda, M. 39
Cazala, P. 16
Chaney, R. B., Jr. 55
Chen, C. S. 22
Chuden, H. 12
Chumak, P. N. 18
Clemis, J. D. 13
Clevenger, W. L. 43
Clow, R. 11
Colburn, H. S. 65
Coleman, H. J. 12
Colman, R. 64
Comber, C. T. T. 40
Conley, F. K. 35
Conley, V. B. 1
Conn, F. W. 7
Connell, J. 43
Conrad, D. W. 23
Conture, E. G. 59
Cooper, M. 32, 61
Corotis, R. B. 61
Cozad, R. L. 59
Cremer, L. 1
Crocker, M. J. 2, 3
Cunningham, M. 57

Dahlstedt, S. 25
Davies, H. G. 47
Davis, M. 57, 58
De Asis Alonso, R. 66
Dear, T. 36
Delmas, A. 17
Dembowski, P. J. 58
Denisov, E. I. 12
Deutsch, C. H. 43

Deutsch, D. 29
Diehl, G. M. 14, 60
Dieroff, H. G. 18, 67
Dillons, J. A. 30
Diserens, A. H. 50
Dittrich, F. J. 45
Djupesland, G. 41
Dokukina, G. A. 65
Donovan, J. 25, 46
Drescher, D. G. 46
Driscoll, D. A. 42
Duifhuis, H. 15
Dunn, S. E. 14
Duvall, A. J., III 62
Dworkin, M. L. 51

Eatock, H. C. 18
Ebbing, C. E. 34
Edwards, D. C. 62
Efron, R. 18
Eleftherious, B. E. 29
Ellingsworth, R. K. 26
Else, D. 48
Emerson, P. D. 9
Enin, I. P. 50
Erickson, M. H. 28
Erulkar, S. D. 18
Es'kov, E. K. 57
Etter, L. E. 51
Evdokimova, I. B. 63
Everest, F. A. 2

Falk, S. A. 46
Fath, J. M. 2
Faulkner, H. B. 17
Feldmann, H. 32
Findley, R. W. 61
Fink, M. R. 65
Fisher, S. 19
Fisk, D. J. 52
Fisk, M. C. 49
Fleszar, I. 16

Florczyk, R. S. 47
Floyd, M. K. 2
Forehand, R. 23
Foreman, J. E. K. 10
Foster, K. R. 39
Fox, J. L. 66
Fraser, J. 32
Freese, A. S. 17
French, A. S. 64
Funasaka, S. 10
Furukawa, T. 63

Gallop, J. C. 46
Gapanovich, V. Ia. 8
Garber, S. F. 24
Garcia-Bengochea, F. 63
Gattoni, F. 7
Geber, W. F. 34
Gel'tishcheva, E. A. 13, 20
Gerver, D. 24
Ghueden, H. G. 37
Gibbs, G. W. 51
Gibson, W. J. K. 67
Glorig, A. 65
Goldberg, S. R. 17
Gol'Dman, E. I. 33
Goldstein, M. 66
Golemba, P. I. 52
Goniashvili, O. Sh. 66
Gottesman, C. A. 30
Gould, W. J. 42
Graevan, D. B. 23
Graham, J. M., Jr. 20
Grant, D. A. 28
Greenwald, A. G. 67
Gregory, F. M. H. 11
Grimes, C. T. 27
Groenwold, F. 38
Gulian, E. 53
Gunasekera, W. S. 34
Gunn, W. J. 32
Guttmacher, H. 31

Hagerty, D. J. 43
Haggerty, H. 59
Hamburg, J. A. 47
Hamernik, R. P. 33, 34
Hammelburg, E. 46
Handel, S. 34
Hankel, K. M. 16
Hankins, W. G. 17
Hanson, D. B. 66
Harmon, B. 57
Harris, R. W. 2
Hartley, L. R. 22, 50
Hauskins, J. B., Jr. 36
Hawkins, J. 49
Hay, I. S. 58
Healey, G. B. 24
Heer, J. E. 21
Hegvold, L. W. 58
Helms, J. 19
Henderson, D. 10, 63
Hennage, J. 42
Henry, K. R. 33
Hersh, A. S. 15

Herzog, R. E. 8, 47
Highstein, S. M. 40
Hiller, U. 40
Hillman, H. 49
Hilton, D. A. 44
Hind, J. E. 50
Hixson, E. L. 38
Hodges, R. 60
Hoefig, W. 60
Holland, F. 46
Holvenstot, C. 18
Hopkins, C. 68
Horch, K. 11
Huang, C. C. 51
Hubner, G. 8
Hughes, L. 48
Humperdinck, K. 33
Hundseth, T. 45

Hunter-Duvar, I. M. 23
Hutt, S. J. 61

Iasinovskii, M. A. 42
Igarashi, M. 28
Ihde, W. M. 19
Il'nitskaia, A. V. 14
Istre, C. O., Jr. 33
Iversen, J. 58

Jackson, R. P. 53
Jacobson, R. A. 28
Jensen, J. 53
Jensen, P. 16
Jerger, J. 6, 9, 18
Jobe, P. C. 21
Jokinen, K. 52
Jongkees, L. B. 68
Jordan, R. E. 31
Jordan, V. M. 13
Jungsbluth, K. 62
Juzwisk, I. 20

Kajlànd, A. R. 38
Kalsi, S. S. 7
Kaltsounis, B. 22
Kastan, D. 47
Katinsky, S. 13
Katsuki, Y. 60
Kavoussi, N. 56
Keating, L. W. 59
Keith, R. W. 24
Kelsey, P. 42, 51
Khaimovich, M. L. 21
Kiang, N. Y. 64
Kiester, E., Jr. 46
Killion, M. C. 48
Kindlmann, P. J. 38
Kinney, W. A. 31
Klepper, D. L. 61
Klingholz, F. 17
Klockhoff, I. 15, 30
Klonowski, S. 14

Klossner, H. O. 11
Klujber, F. 18
Koch, U. 29
Kodaras, M. 15
Kodaras, M. J. 60
Koller, M. 5
Koltzsch, P. 53
Kondo, S. 64
Kowalewaska, M. 11
Kramarenko, I. B. 61
Krantz, D. S. 31
Krell, D. 7
Krochmalska, E. 21
Krokosky, E. M. 43
Kruzhkova, G. V. 18
Kurze, U. J. 47

Labiak, J. M. 20
Laciak, J. 23
Laird, C. W. 11
Landwehr, F. J. 10, 55
Lane, E. 60
Lane, S. R. 30, 35, 39, 56
Lang, W. W. 46
Langdon, L. E. 35
Lapchuk, T. V. 59
Lass, N. J. 48
Latkowski, B. 27
Lefaucher, C. 67
Leigh, J. M. 9
Lenhardt, M. L. 23
Letourneau, J. E. 22
Levere, T. E. 9
Liebig, J. 2
Liebig, W. 51
Lilley, G. M. 45
Ling, D. 67
Lipscomb, D. M. 2
Livesey, D. L. 27
Lockett, M. F. 59
Lou, S. C. 44
Loucks, D. P. 26
Lowery, R. L. 44

171

Luz, G. A. 55, 63
Lybarger, S. F. 6
Lyon, R. H. 2, 57
Lyon, T. A. 29

Maass, B. 51
MacDonald, K. 31
Macduff, J. N. 10
Macrae, J. H. 45
Mac Whorter, R. F. 29
Magie, A. 54
Maglieri, D. J. 45
Maksimova, L. I. 21, 59
Malinowski, T. 38
Malyshev, E. N. 21
Mann, A. 22
Marcus, A. H. 64
Marshall, L. 64
Martenhoff, J. 60
Martin, A. M. 9, 39
Marwood, J. F. 36
Mast, T. E. 19
Masterton, R. B. 7
Matsnev, Z. I. 28
McCallum, P. 38
McCoyd, K. 5
McGrath, D. C., Jr. 67
McGuinness, D. 26
McRandle, C. C. 20
Meier-Ewert, K. 5
Meisner, D. 41
Melnick, W. 33
Mendel, M. I. 15
Mensikova, Z. 5
Merluzzi, F. 25
Merriman, R. M. 20
Meteer, C. L. 62
Metlyaev, G. N. 48
Meyer, A. F., Jr. 50, 67
Meyer-Rochow, V. B. 59
Michel, R. C. 29
Middleton, D. 37

Miezejeski, C. M. 22
Miller, E. F., II 61
Miller, J. D. 24
Miller, M. H., Jr. 42
Miller, R. K. 46, 55
Mills, J. H. 17, 65
Mis, F. W. 35
Miska, K. H. 28
Mitchell, E. E. 6
Mitsuya, T. 18
Miura, H. 62
Moch, P. 33
Molino, J. A. 26
Montandon, A. 18
Moody, R. 60
Moore, B. C. 59
Moran, R. D. 52
Mordey, R. 50
Morgan, D. E. 36
Moriyama, H. 42
Morley, B. J. 31
Moroz, B. S. 8
Morrow, C. T. 44
Mountjoy, J. R. 41
Moushegian, G. 36, 55
Muller, G. 6
Mulroy, M. J. 35
Murata, K. 10
Murray, R. 38
Murtland, W. 16
Musa, R. S. 15
Muzet, A. G. 11

Nabelek, A. K. 40, 55
Nader, R. 54
Nagel, D. C. 48
Naumowski, Z. 54
Nekhorosheva, M. A. 13
Nelkin, D. 38
Neville, H. 25
Newman, H. 54
Nezer, Y. 41

Nolen, W. A. 32
Norman, R. S. 52
Novak, E. 8
Novak, R. 23
Novick, S. 35
Nuenke, R. H. 61
Nugent, R. 16

Odkvist, L. M. 67
Okitsu, T. 10
Oleshkevich, L. A. 14, 22
Oliphant, K. S. 12
Olsen, W. O. 11
Olton, D. S. 57
Oppenheimer, R. P. 30
Orlovskaia, E. P. 33
Ornitz, E. M. 55
Ortiz, G. A. 40
Osada, Y. 50
Osman, A. E. M. 43
Ouellette, P. L. 65
Oyama, H. 56
Ozerengin, M. F. 26
Ozsahinoglu, C. 63

Paran'ko, N. M. 47
Parlin, D. B. 6
Pearlstine, N. 50
Pedersen, C. B. 37
Pepersack, J. P. 15
Perrin, R. G. 6
Petrovi, D. 43
Petrusewicz, S. A. 2
Pfaltz, C. R. 62
Phillips, W. G. 17
Piazza, R. 37
Picar, P. 67
Pickles, J. O. 57
Pierson, W. R. 43
Pietruck, S. 43
Pinder, C. A. 47
Pinheiro, M. 56

Plett, E. G. 35
Pollitt, W. A. 45
Popper, A. N. 56
Poulton, E. C. 34
Powell, C. A. 24
Powell, M. 12
Price, G. R. 36, 65
Pringle, J. W. 58
Pugh, J. E., Jr. 13
Pulec, J. L. 36
Pye, A. 6

Quante, M. 21

Ragan, C. A., Jr. 69
Rakhmilevich, A. G. 61
Rastorguev, O. S. 39
Ratner, M. V. 29
Raushenbahk, I. Y. 40
Ravizza, R. 57
Reese, J. A. 11
Reshotko, M. 25
Ringham, R. F. 65
Robertson, D. 13
Robertson, R. M. 21
Rodenburg, M. 8
Roedler, F. 36
Romand, R. 52
Ronken, D. A. 37
Rosen, G. 10, 53
Rosengren, L. G. 63
Rosenhouse, L. 42
Rouse, W. H. 68
Rudenko, V. F. 34
Rudinsky, J. A. 60

Sachs, M. B. 19
Sager, O. 56
Sampere, J. M. 29
Sanders, J. W. 9
Sanders, M. 20
Sataloff, J. 3, 52

Sato, S. 62
Saunders, J. 46
Saunders, J. C. 63
Schafer, R. 3
Scheiblechner, H. 67
Scheuch, K. 13
Schiff, M. 48
Schindler, R. A. 13
Schlicke, H. M. 30
Schmidek, M. E. 24
Schmidt, C. 65
Schmitz, F. H. 55
Schmitz, H. 60
Schneider, C. E. 28
Schnieder, E. A. 16
Schreiner, L. 47
Schroeder, M. R. 39
Schultz, T. J. 49
Schwartz, A. H. 67
Schwartz, E. 36
Schwartz, S. 9
Schwetz, F. 9
Scott, C. E., III 18
Secretan, J. P. 38
Semjen, A. 35
Sharp, B. H. 27
Shirreffs, J. H. 55
Shuey, S. J. 68
Shumann, W. A. 7, 27, 35
Siddon, T. E. 37
Siegelaub, A. B. 31
Silverstein, H. 63
Silvidi, A. A. 46
Simacek, I. 7
Simmons, F. B. 24
Slob, A. 20
Smith, H. G. 14
Smookler, H. H. 33
Snowden, J. C. 2
Sobolevskaia, L. A. 44
Soderman, P. 50
Sohmer, H. 25

Soldatkina, S. A. 17
Sondhi, M. M. 40
Sonn, M. 24
Sova, R. E. 8
Speaks, C. 34
Spellenberg, S. 67
Stafford, J. R. 3
Starr, A. 41
Stawinski, S. 21
Stefenov, B. 2
Stelmack, R. M. 27
Stephens, R. W. B. 3
Stephenson, J. E. 22
Stepniewski, W. Z. 52
Sterkers, J. M. 20, 39
Stevens, M. W. 7
Stitt, C. L. 40
Stokinger, T. E. 37
Storrs, L. A. 5
Strasser, H. 14
Stuck, B. W. 62
Sturges, P. T. 56
Sturzebecher, E. 62
Suga, F. 29
Suga, N. 8
Suladze, E. Sh. 40
Sulkowski, W. 21
Summar, M. T. 33
Sundberg, J. 9
Sutherland, L. C. 59
Suzumura, N. 36
Swaan, D. 10
Swing, J. W. 39

Tabaczuk, E. 49
Takahashi, M. 8
Tammen, H. 33
Tarmas, J. 40
Tatusesco, D. 3
Taylor, E. M. 7
Taylor, G. E. 29
Taylor, R. 45

Taylor, R. L. 8
Tew, J. M., Jr. 20
Thackray, R. I. 50
Thayer, R. E. 60
Theiss, W. R. 33
Theologus, G. C. 23
Thomsen, J. 66
Thumann, A. 34
Tipton, A. 68
Tirumalese, D. 20
Tokarev, V. A. 26
Townsend, T. H. 31
Train, R. E. 45
Tunner, J. V. 67
Turkewitz, G. 56
Tyler, D. A. 43
Tyler, J. M. 61

Ulrich, R. F. 64
Unger, H. G. 48
Utidjian, H. M. 17

Vanderhei, S. L. 22
Vasil'eva, A. L. 30
Veidaver, H. 31
Vlasenko, V. V. 30
Volpliushkin, E. I. 5

Wagner, G. 14
Wagner, H. 53
Wajskop, M. 39
Wakefield, B. D. 38, 39, 42
Wakeley, J. 63
Walloch, R. A. 51
Wanke, E. 68
Warnock, A. C. C. 6
Watson, J. E. 47
Weinstein, N. D. 22
Weissenburger, T. T. 34
Welch, B. L. 50
Wells, A. 48
Wende, E. 37

Wende, S. 64
Wesler, J. E. 14, 63
Whitcomb, C. E. 62
Whitfield, I. C. 52
Whyte, B. 12
Wick, C. H. 16
Wiggins, J. H. 4
Wightman, F. L. 51
Willott, J. F. 10, 53
Willoughby, R. A. 41
Wilson, C. E. 44
Wilson, D. 44
Winnerling, H. A. 54
Wiss, J. F. 68
Wojda, H. W. 62
Wolcott, G. T. 39
Wolff, P. H. 22
Wright, J. W. 5
Wright, W. M. 19

Yaffee, M. L. 41
Yamamura, K. 24
Yamazaki, M. 41
Yerger, L. F. 38
Yoshimoto, Y. 18
Young, E. 55
Young, G. C. 15

Zamfir, G. 57
Zaslavskii, I. E. 6, 27
Zekem, H. B. 24
Zippel, U. 13
Zulfacar, A. 31

Ref
Z
5862.2
N6
B55
1974